Table of Contents

INTRODUCTION

Anyone can draw but not everyone can make good drawings. Pencil drawing is a skill that needs a good foundation on theories because pencil drawing is a blend of theory and proper execution of these theories.

Practice makes drawing perfect but foundation on the theories and techniques in drawing paves the way for better drawings.

This ebook mainly targets beginners in pencil drawing and those who wish to enhance their pencil drawing skills through other techniques and insights that one may find here.

Anyone may use this ebook to hone his or her drawing skills with the end goal of becoming a better pencil-drawing artist through this small tribute.

This ebook, written in non-technical language, seeks to promote better understanding. It covers the basics of pencil drawing, providing for a good foundation for pencil drawing and some practical tips.

This ebook contains articles on pencil drawing, and it shows how to draw step-by-step common objects such as people, cars and animals.

Since drawing is visual, chapters are short providing for more illustrations and application of drawing theories.

The maker of this ebook hopes to inspire individuals to pursue the art of pencil drawing and unleash their artistic mind.

What it takes to be a Good Artist

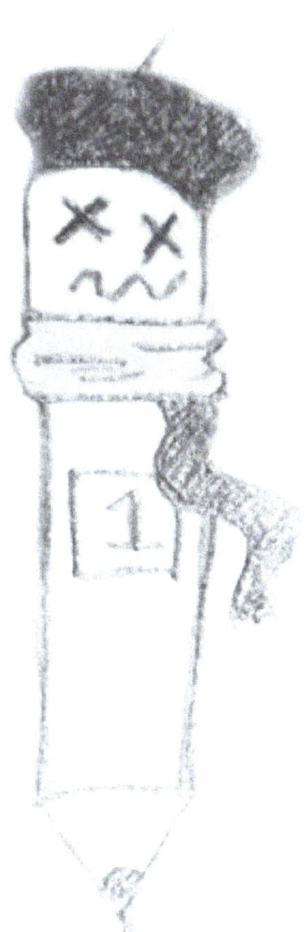

Artists are made not born. Unlike medicine or law that one has to have proper education to be equipped with the degree of knowledge, most of the artists around today are self-taught artists. Going to a formal art school is not a pre-requisite before one becomes an artist. Unless if one desires to be a commercial artist or one that works for companies or newspapers.

The artist's journey starts with the desire to express themselves through drawing and art. In fact, they can express themselves better than they do with words. Drawing comes naturally for them.

Some could express themselves through cartoons or comic strips while others make abstract paintings. Just because drawing comes naturally for them, it doesn't mean that they dealt away with the basics of drawing. They still learned the basics of drawing, but they adopted their own style in drawing.

Having technical skills in drawing is good but drawing should be more on self expression. In drawing, there is no correct or wrong style because drawing is self expression. It is up to the viewer to appreciate that particular self expression or not.

These artists were curious and discerning about the world around them. They have an eye for details and how objects and scenery were composed by nature. Having the drive and persistence to imitate nature and give life to their vision, artists constantly practice to perfect their craft. Deciding to become an artist is an easy decision but becoming one is not quite as such.

Advantages and Disadvantages of Pencil Drawing

The advantage of using a pencil when drawing as opposed to using a pen is that you can easily erase mistakes when you use a pencil for drawing. Artists can make mistakes and not have to start all over again. Erasing is not just for mistakes.

Pencil drawing is a process, artists start drawing by making light outlines that help them create a drawing.

You can also erase later on the outlines and people will hardly notice that the drawing came from simple lines. Using pencils in drawing is inexpensive because you will just need a pencil and paper to create a basic drawing.

Pencils and papers are easy to carry around that you do not have to confine yourselves to your studio when making a drawing.

Pencils are affordable and sold in most stores. You do not have to worry about the pencil drying up or reaching its expiry date. It can be stored anywhere and maintaining it is hassle free.

Unlike using paints and other media that give off nasty fumes, pencils are odorless. Manufacturers made pencils versatile depending on the artist's needs, one get very black to light gray. Pencils can also be very soft, which makes it easy for shading or hard, which makes it perfect for fine details.

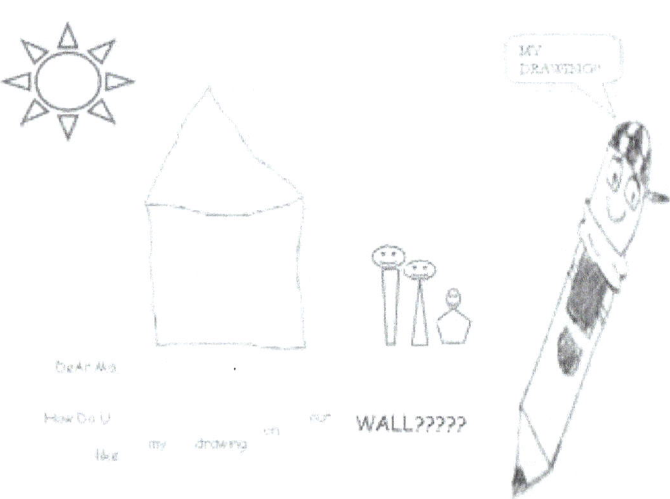

Pencil drawings have a tendency to smudge if you touch or rub them. Those once fine details will look like a smear. It is important to keep hands away from the paper while drawing or cover your drawing with another piece of paper while trying to complete the work.

Once the drawing is completed, seal the drawing to prevent from smudging. Sealing is also important because of the temporary nature of pencil drawings, and you can use a sealing or a fixative spray for this. Others use hair sprays that are perfume free. Artists lightly spray on these on their work.

BRIEF HISTORY OF PENCIL DRAWING

Famous Pencil Drawing Artists and their Works

Using a pencil has many advantages, one of which is that it is very easy to erase one's mistakes. There are famous people in the world who preferred to use pencils.

Vincent van Gogh, a notable artist used Faber-Castell pencils. According to him, these pencils were superior. He liked its blackness and found it easy to use.

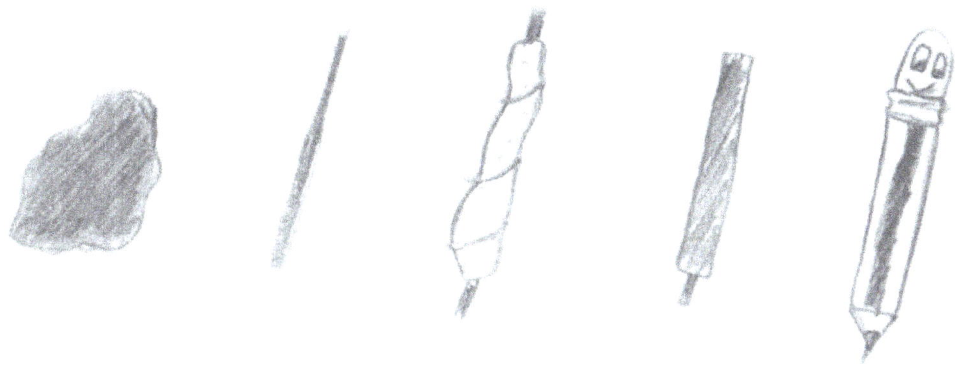

Pablo Picasso used a pencil with his works of art. Jean Auguste- Ingres made the Portrait of Mme Guillaume Guillon Lethiere. John Constable used a pencil and a sepia wash in his Trees and a Stretch of Water on the Stour.

These works are in different famous art galleries and museums around the world.

The Story of the Pencil

Long time ago, Greek artists used to draw on papyrus using a metal stylus. In Cumbria England, there were large deposits of graphite. Locals used the graphite for marking sheep. Although graphite was solid, it was soft that it needed a case before you can use it easily.

Early civilization used sheepskin to wrap the graphite. England continued to produce pencils while in 1662; Nuremberg Germany attempted to manufacture pencils using a mixture of graphite, antimony and sulphur. Italy was the first to use wooden holders.

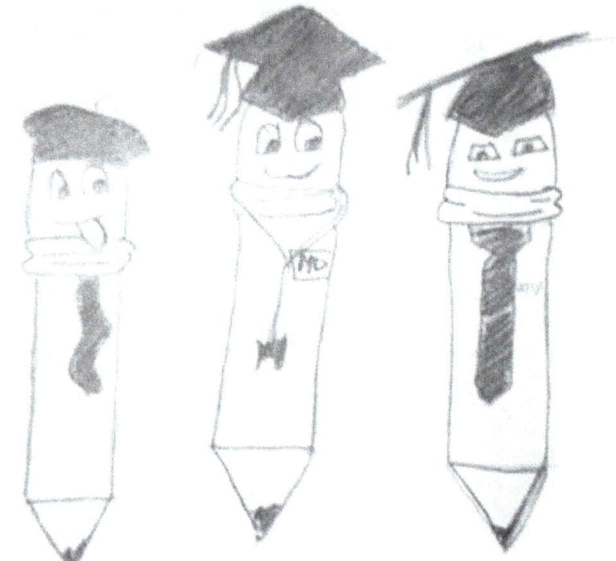

They carved two wooden halves and inserted the graphite stick. The wooden halves were then glued together.

Many years later, Ebenezer Wood in America automated the process of pencil making. Joseph Dixon found a way to mass produce pencils. Hymen Lipman attached the eraser to the pencil.

Today, annual production of a pencil is at around 14 billion pencils.

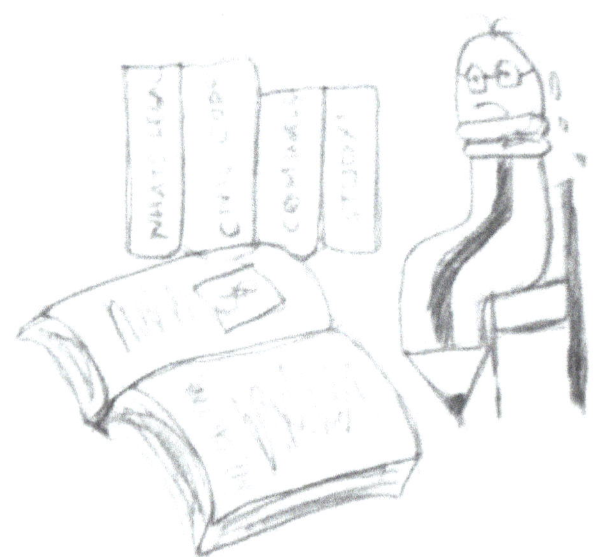

These pencils may be round-shaped, triangular or hexagonal. There are even bendable pencils. Pencil variations include mechanical pencils, which has a mechanism that pushes the lead through the end. A Quadra chromic pencil has four colors at its tip.

GETTING STARTED

Materials and Tools:

Choosing the Right Kind and Quality

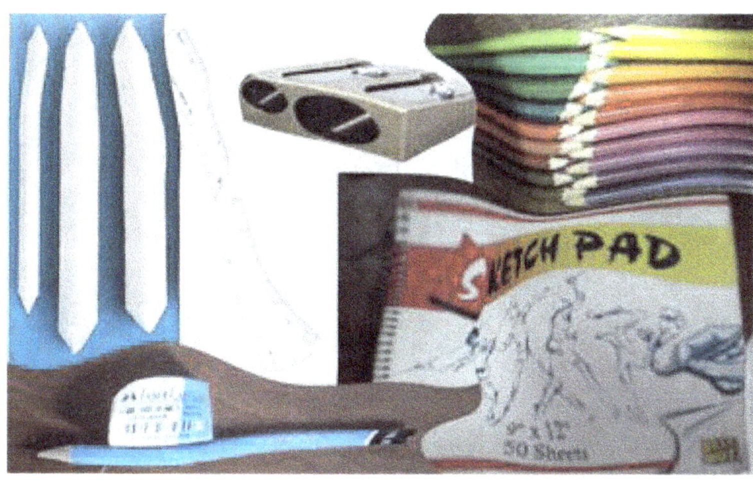

The age old debate about the artist's skills' versus high quality materials is still applicable even until today. Quality of tools and materials used affects the artwork.

One cannot aspire to own and use top quality art materials, especially when one is just beginning. The best attitude towards the problem of not having enough materials is this:

It is more miserable to have no skills in art but have the best materials and tools available than to have mediocre materials but have excellent artistic skills.

Pencil

The pencil is an instrument commonly used to drawing. A stick of graphite or a mixture is encased in wood. At one end of the pencil is an eraser. There are many kinds of pencils.

Manufacturers use "H" for hardness, "B" for blackness, "F" for a fine point. Pencils can be very dark or very light. Artists may need different pencils for making grays and blacks.

Pencils can also be hard or soft. Hard pencils are needed by engineers because it gives them better control in making precise lines. For users who cannot decide which one to use, the safest is HB. Some artists use the non photo blue pencil. It is a pencil used mainly for outlining. It comes in a very light blue color, a shade that cameras cannot catch. Artists can save time with this pencil because they do not have to erase it.

Mechanical pencils are available in .5mm or .7mm lead. Mechanical or automatic pencils do away with the need to sharpen the pencils. The downside of using mechanical pencils is the lack of variety of grades.

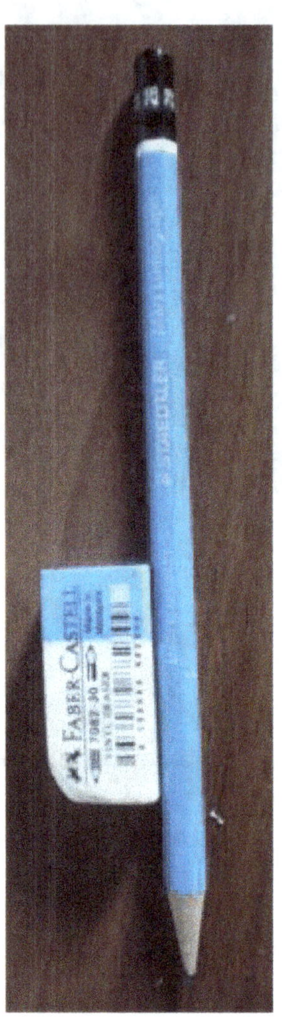

Eraser

It is impossible not to make a mistake while drawing. More often than not, one consumes the eraser at the tip of the pencil before the pencil. An eraser is an essential tool in an artist's toolbox.

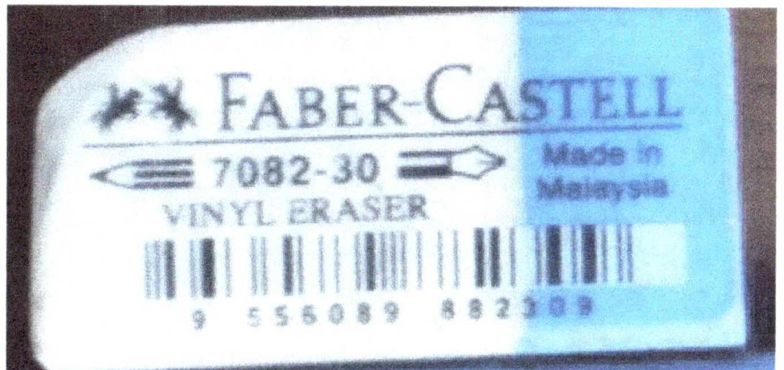

Erasers may come from pulverized pumice. Do not use this type of eraser to heavily, or it will damage the paper. After erasing, brush away the residue that this type or eraser will leave.

Soft vinyl erasers are softer and cause lesser damage on paper than erasers made from pumice. This eraser is good for precision erasing and erasing light marks.

The kneaded eraser reminds people of chewing gum. Unlike the eraser made from pumice, this type of eraser does not leave any residue. With the use of your fingers, you can shape kneaded erasers making it easy to erase little details. This eraser is difficult to use in large areas.

An art gum eraser is another type of eraser that is widely used among artists. Unlike the kneaded eraser, removing large areas is easier with this eraser. The downside of using a gum eraser is that it is not precise in erasing fine mistakes.

Drawing Pad

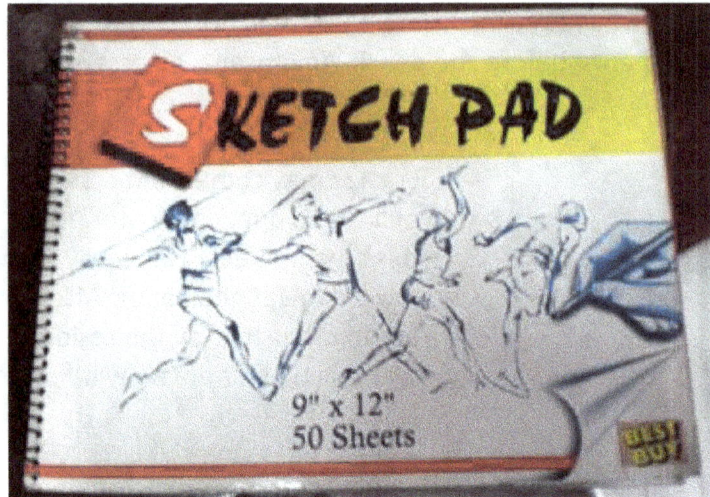

A drawing pad is also known as a sketch pad. It is made up white paper bounded in one book. It can be used for drawing and scrap booking. Drawing pads come in different sizes. It's always important to have a sketch pad to contain all your drawings.

Practice bringing a small sketch pad around, so that you can easily draw what captures your attention. Leonardo da Vinci always had a sketch pad or a notebook with him. It allowed him to draw everyday objects and movement with ease.

Drawing board

A drawing board can be used for several purposes. It can be used for writing, reading large documents and making large sketches. This equipment is highly optional, unless you aim to draw in large scale.

Paper Stumps or Cone Blenders

Blending is a big part in drawing. Do not use fingers because fingers leave oil on the paper which can be visible in certain light. Artists use blending stumps or paper stumps to blend large areas.

Pencil Sharpener

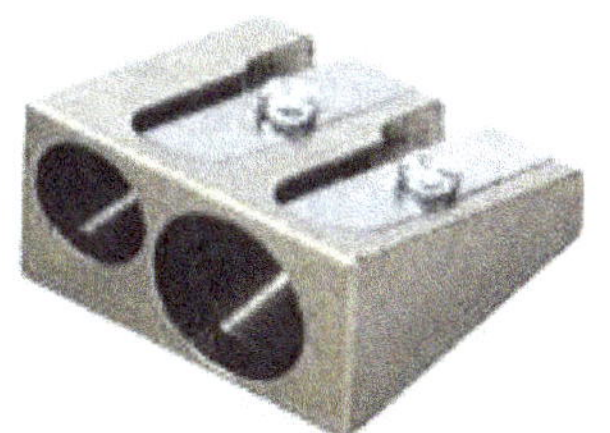

Another essential tool that should be found in the artist's toolbox is the pencil sharpener. This is a device that shaves the end of the pencil thereby sharpening its point.

Some artists may use sandpaper block and use a knife for the wooden part of the pencil. A large part of pencil drawing depends on keeping a sharp pencil.

Ruler

A ruler is used in technical drawing, engineering and geometry. Rulers help you make straight lines or make precise measurements. On the topic of perspective, the ruler will be frequently used.

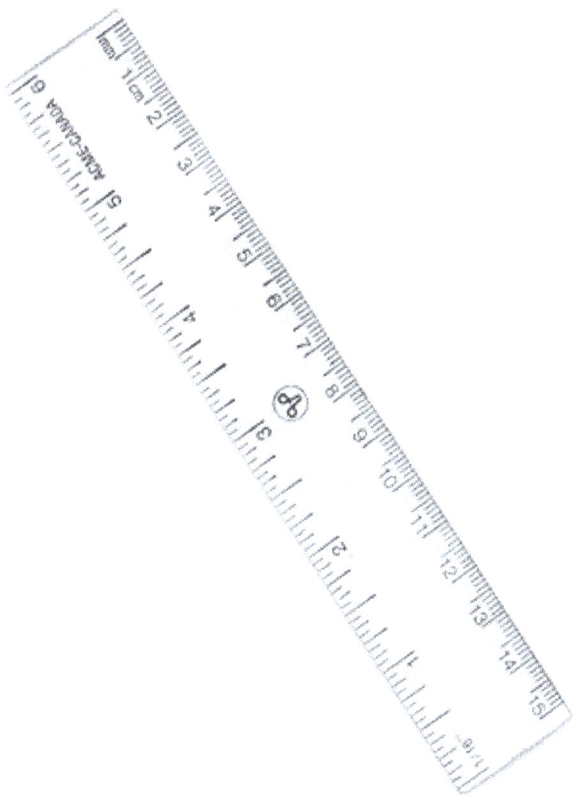

LEARNING THE BASICS IN DRAWING AND SKETCHING

How to Hold the Pencil? - The Different Pencil Grips

Tripod Grip

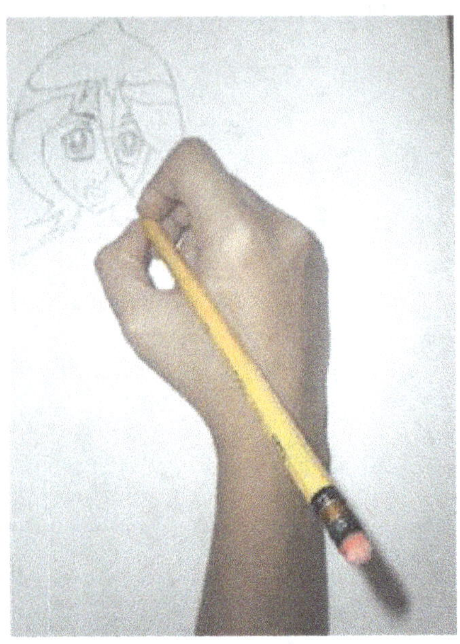

The tripod grip is the most common way of holding a pen or a pencil. Although it might be the common way of holding a pencil, many people are not doing the tripod grip correctly. The correct way to make a tripod grip is to position the pencil applying equal pressure between the side of the middle finger, the tip of the index finger and the thumb.

Do not grip the pencil too firmly. Holding a pencil too tightly or too lightly will limit your flexibility in drawing because the fine-motor skills are weakened.

The index finger rests on the tip of the pencil which shows the lead while the tip with the eraser must be pointing toward the shoulder.

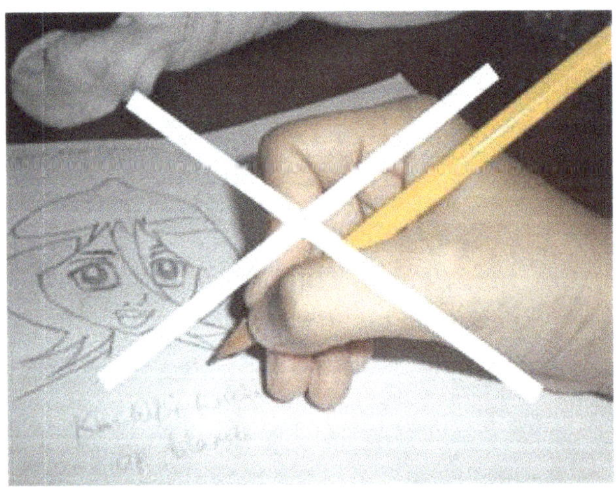

Do not hold the pencil vertically. Check if the thumb and the underside of the forearm are in a straight line.

The most important and apparent advantage of this type of grip is that a person can write more quickly and easily. The correct grip maximizes one's motor skills, which are needed for drawings. Also, it reduces the risk of physical problems, especially the carpal tunnel syndrome.

Extended Grip

The extended tripod grip is a variation of the tripod grip. The tripod grip is still used, but you will hold the pencil at bit further at the end or near the tip that has the eraser. Because your control is at the other end of the pencil, small movements of your hand make larger effects on the other end.

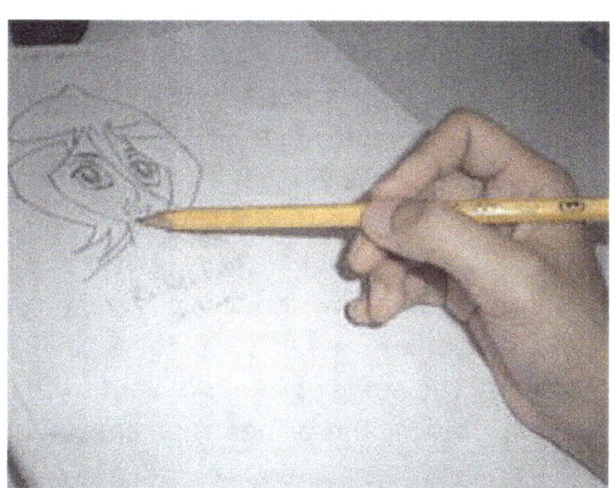

Underhand Grip

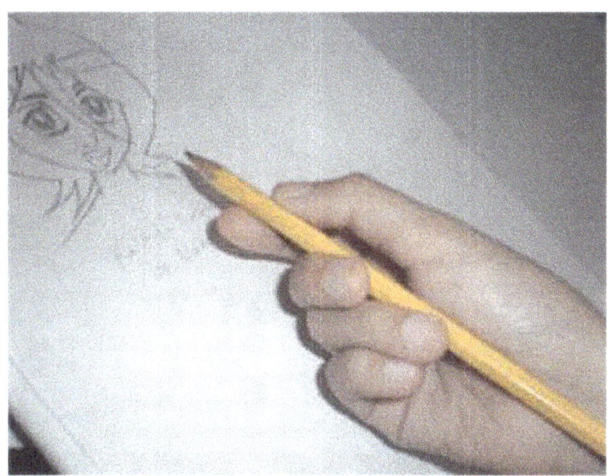

Another variation of the tripod grip is the underhand grip, there is also an underhand grip. This is another relaxed way of holding a pencil. This grip is best suited for broad sketching. The pencil is positioned in a 'V' with the middle and index finger controlling the movement. Using this type of grip will help you make firm lines and small linear details.

Overhand Grip

For sketching, most artists use the overhand grip. The aim of the overhand grip is to have a relaxed grip on the pencil but not too relaxed that the grip would not be secure.

You can draw sitting or standing, make sure that your arm has full range movement. Shading is easier with the overhand grip.

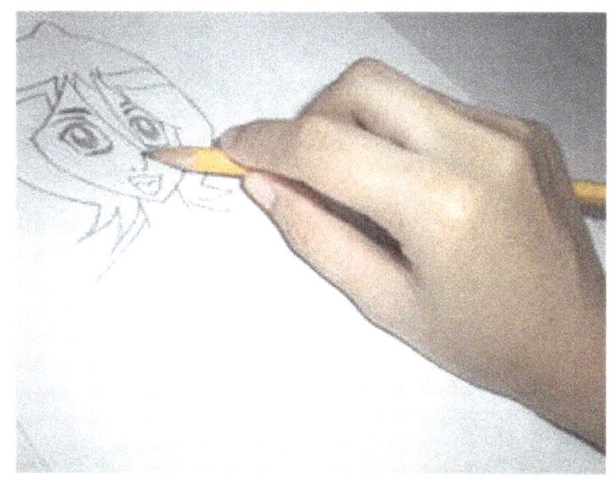

How to Draw Lines?

In truth, lines do not exist and those that you see as lines are but your observations of an object's edge. Lines are indeed man's creation used to represent and mimic the objects and figures that he sees. It is a basic element of art that you would never conceive a single work of art without the application of lines. Lines enable artists to create both space and image.

There are so many kinds of lines and these lines vary in terms of length, width, value and in many other ways. In form, there can be straight, curved, wavy, jagged and many others. The range of values of lines from light to dark is a product of pressure placed while drawing these lines. Here are some of the basic lines used in pencil drawing.

Flat Lines

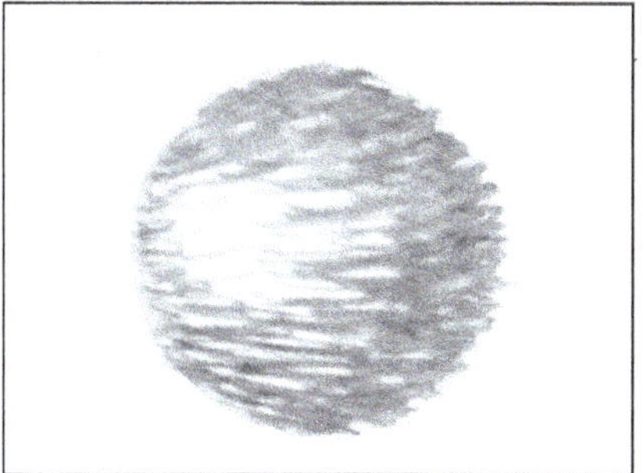

Horizontal lines

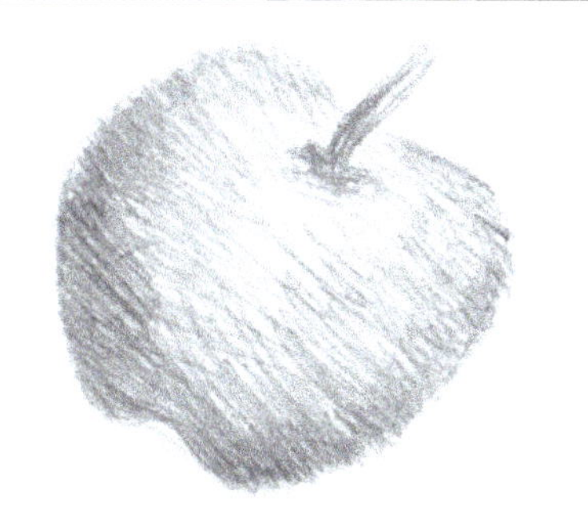

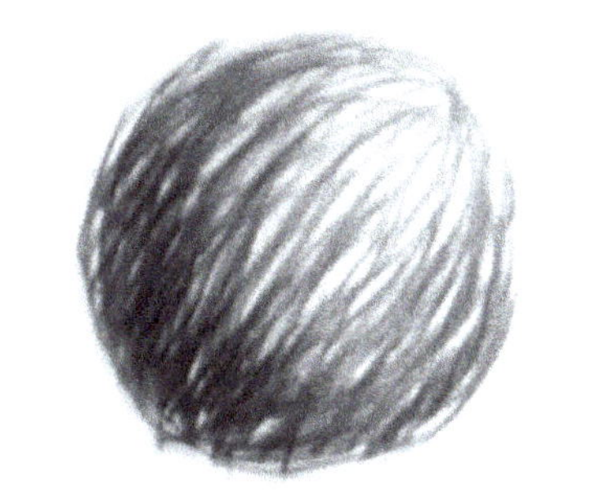

Vertical lines

Flat lines, also commonly known as the straight lines, are helpful in expressing emotional states and evoking emotional responses and providing illusion on the viewers' eyes.

You can start on a horizontal direction, a vertical direction, or a diagonal direction depending on how you want your image to look like.

Generally, the horizontal approach gives your image a wider look while the vertical direction gives your image a leaner and thinner look.

In terms of emotions, horizontal lines invoke serenity and stability and vertical lines invokes poise and stillness.

In fact, artists employ vertical lines to show strength and dignity with the use of height often reflected through vertical lines.

Diagonal lines, on the other hand, invoke movement, unrest, change, instability and variation.

Accent Lines

Accent lines find significance in putting certain emphasis or accent in some portions of your drawing. It, specifically gives special emphasis on your certain portion through changes and variations in your lines.

Contour Lines

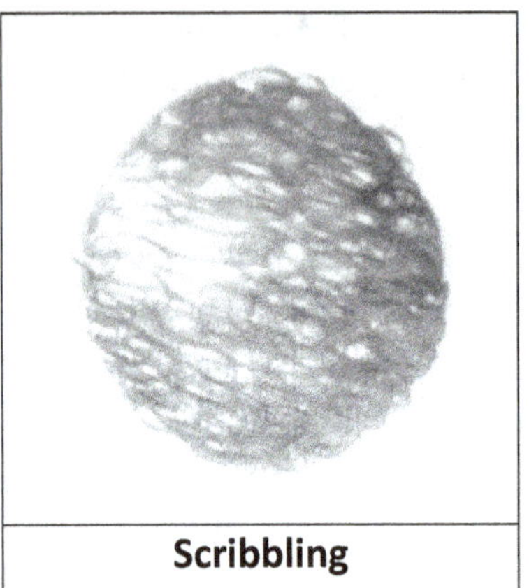

Scribbling

One of the striking characteristics of contour lines is the purity of the line, which even without the use of color, distinctively display shape and beauty of the image. In outlining and in sketching, what you follow is only the important edges, outlines and contours of the object you want to draw.

Scumble/Scribbling

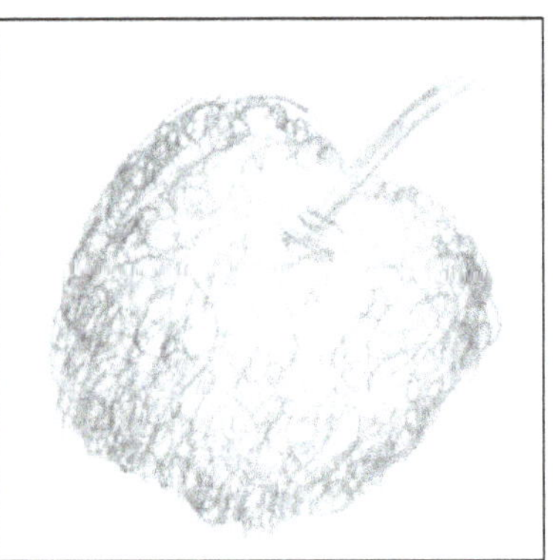

Scribbling or scumble drawing is a drawing merely composed of abstract and random lines.

Cross Hatch Line

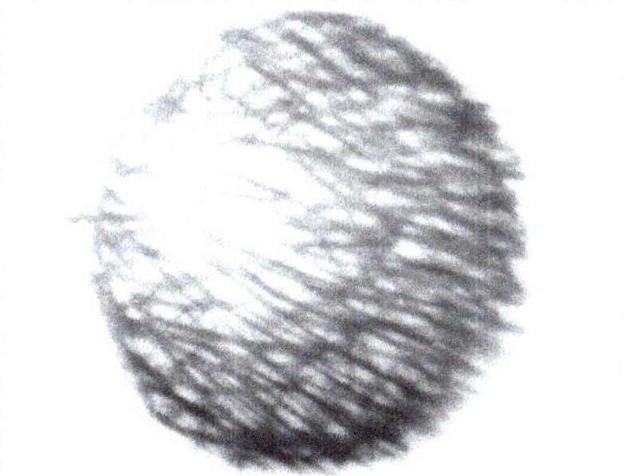

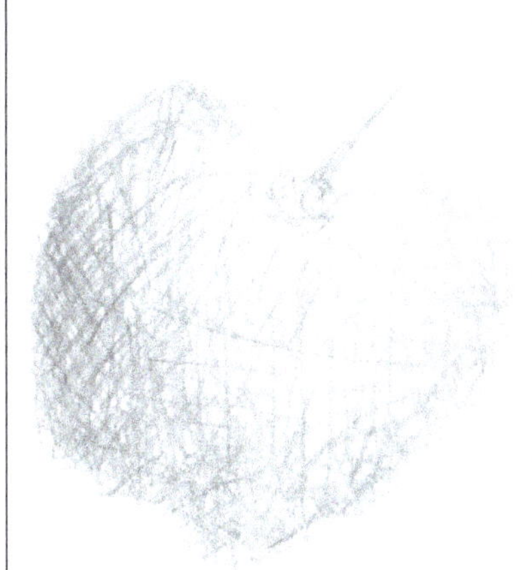

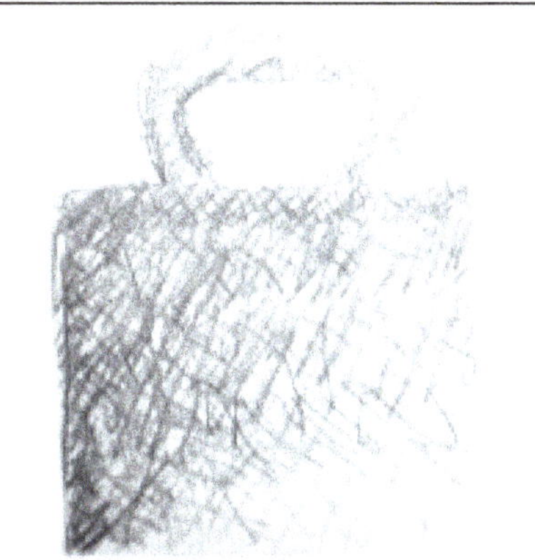

Another kind of line often used in cartoon works is crosshatching.

The shading technique is often different from the lines used in making the drawing but certain cartoonists make use of the lines in creating shading effect.

With crosshatching, interlacing or interweaving parallel lines often in different densities are able to create the effects of light and shadows in their drawings thereby creating shading on the images.

Another application of cross hatch lines is in creating lineal definition and texture on the drawing.

Most drawings of walls, fabrics and other textured materials have the application of crosshatching technique in drawing the lines.

Parallel crosshatched lines provide a flat image while the use of curved lines creates an illusion of plumpness and mass.

Thus, two similar objects may have different illusions due to the use of lines.

Smudge

Smudging is a drawing technique dealing with proper blending. Smudging is easy to do; you simply draw an object and start shading it with the kind of value you want it to have and blending it properly with the use of a tissue, cotton puffs or paper towel.

Pointillism

Pointillism known otherwise as divisionism or stippling is a technique of grouping together dots to create an image or filling an image using dots.

The rule of the thumb is that the closer that you position the dots, the darker that portion that is in your drawing. Pointillists do the reverse by drawing the foreground before filling the inside.

Pointillism is a very slow form of art because technically speaking, filling an image with the use of dots is a very monotonous process.

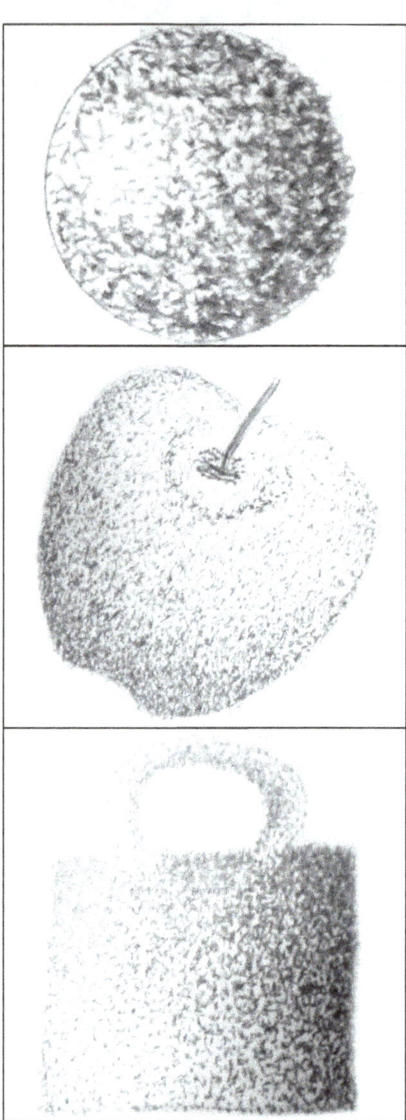

Basic Perspectives in Drawing

An Introduction on Perspectives

One can see realism in a drawing with the application of perspectives in drawing. The idea or concept of perspectives came with the idea of having sets of images that appear coherent and consistent in one piece of art.

Linear Perspective

Linear perspective is a realistic method of portraying objects with a visual depth. The rule of the thumb though is that the farther the object is, the smaller it becomes and conversely, the nearer the object is, the bigger it becomes.

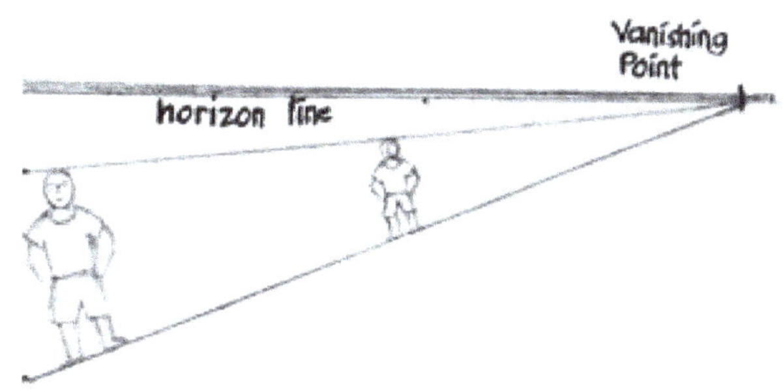

In linear perspective, there are parallel lines receding into a particular distance where the lines seem to converge until these parallel lines meet at a point in a horizon where they disappear, called the vanishing point.

Zero Point Perspective

Since vanishing points exist only with the presence of parallel lines, then for non-linear illustrations where there are parallel lines in the scene, then there could be no vanishing points and thus called zero perspective drawings. Most natural scenes like a range of mountains are typical illustrations of zero perspective drawings though these drawings may still create an illusion of depth on the viewer's eyes.

One Point Perspective

One-point perspective drawings refer to those with only one vanishing point that are often than not directly opposite to the eye of the viewer. In one-point perspective, parallel lines, parallel to the eye of the viewer, retreat towards a certain point in the space called a vanishing point. This often illustrated by illustrations of roads and railway tracks, and hallways that recede at a vanishing point as illustrated.

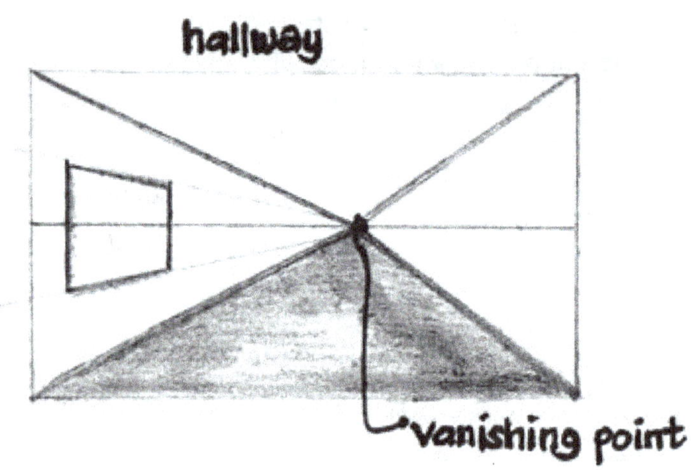

Sample drawing of a box with the use of one point perspective.

1. Draw a horizon line or that line in the plane where the sky meets the earth.

2. From your horizon line, choose a vanishing point that can be the center, near right or near left.

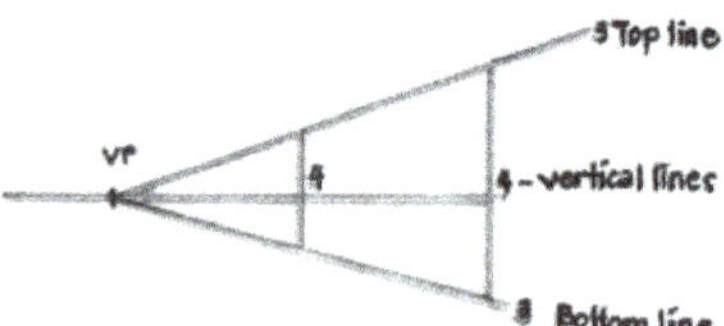

3. From your vanishing point, draw one top line and one bottom line with the top line much longer than your bottom line and this would create the illusion of depth later on. These lines will form the sides of the box.

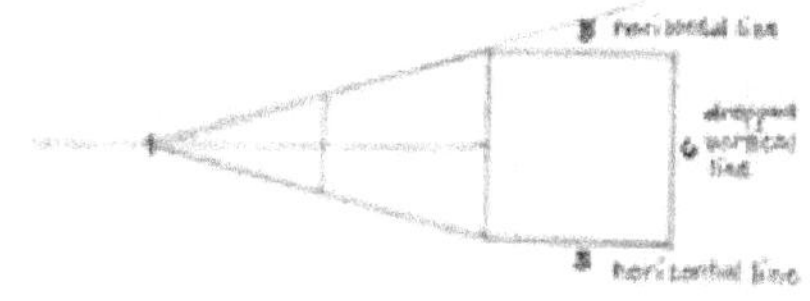

4. Draw two vertical lines that would connect your top line and your bottom line.

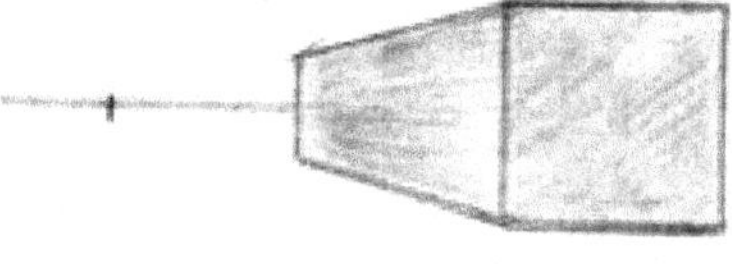

One - point perspective box

5. Then draw the front of the box by drawing two horizontal lines of same length from the top to the bottom to the closest wall to it.

6. Then connect these two horizontal lines by dropping a vertical line on the other side.

Two-Point Perspective

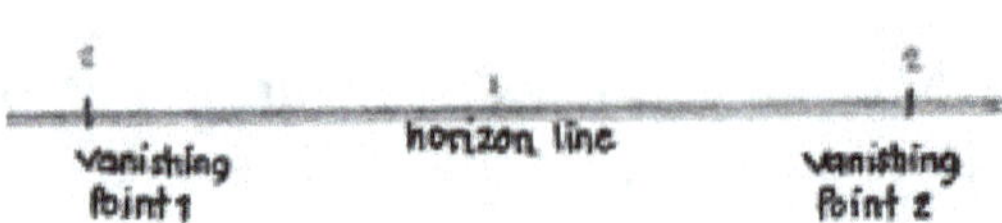

1. Draw your horizon line on the plane.

2. Instead of one point, mark two points, which will serve as your vanishing points on the horizon line preferably on the right and the left side of the horizon line.

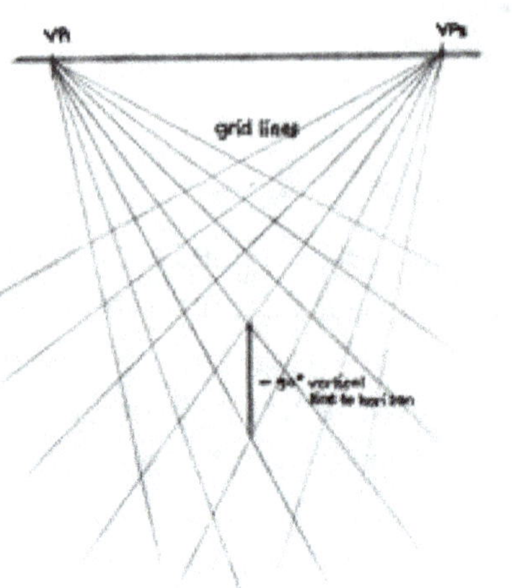

3. Draw your construction grid by finding below your horizon line the middle. The middle is the point where all the lines coming from the vanishing points meet.

4. Draw six lines coming from each of the vanishing points passing the middle.

5. Draw your middle vertical line that is at 90-degree angle with your horizon line, which will serve as the corner of your box.

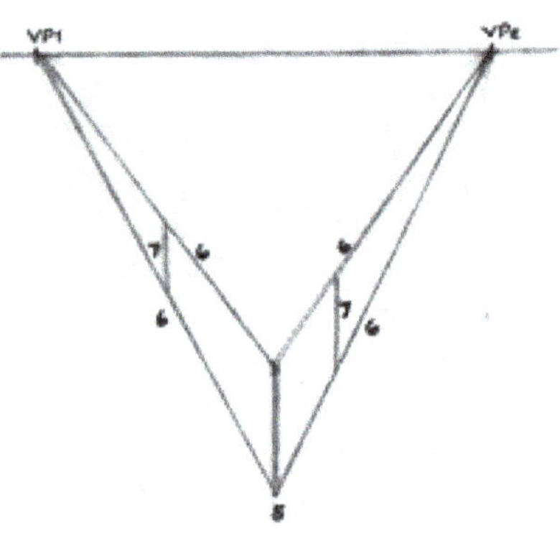

6. Draw the top and the bottom lines connecting your middle vertical line to each of your vanishing points.

7. From your middle vertical line, draw one vertical line to the right and another to the left depending on how you want your box to look, as these would determine the length and width of your box.

8. From your vanishing points, draw the line connecting your right vanishing point to the left vertical line and your left vanishing point to the right vertical line.

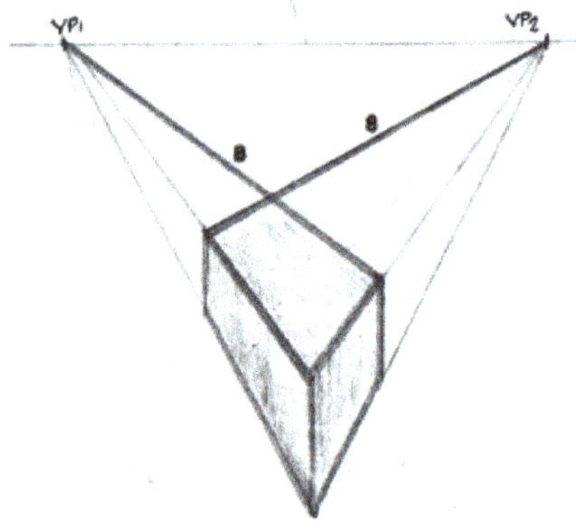

By now, you figure out the two-point perspective box. You may now start shading your two-point perspective box and erase grid lines passing inside the box.

Two - Point Perspective Box

Three Point Perspective

Three-Point Perspective is often used in architectural drawing using lines parallel to the Cartesian axes x, y, and z. This perspective finds relevance in drawing buildings.

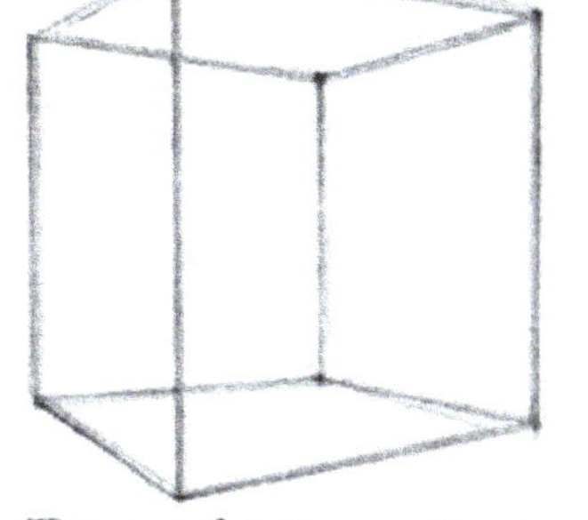

Three-point Perspective Box

Isometric Perspective

This refers to a perspective where distant objects and figures become smaller but the parallel lines do not converge. What happens to be that lines that are perpendicular to the plane become sharp diagonals as drawn thus turning the squares or rectangles into parallelogram.

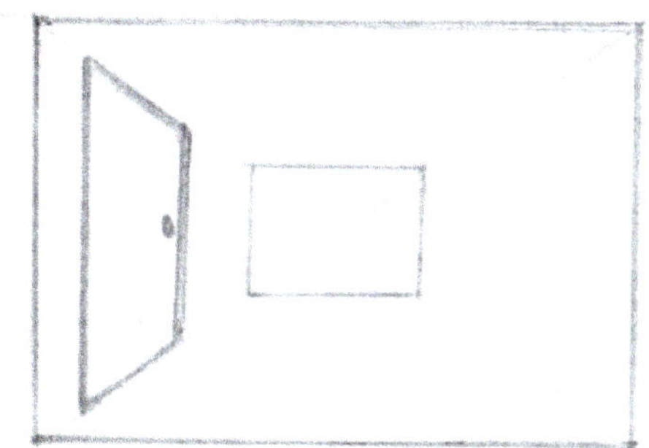

Atmospheric Perspective

Atmospheric perspective or aerial perspective expresses nature in their distance. It simply stated, objects perceived to often appear as blurred, misty looking, and in distinctive or hazy looking to the eye.

As shown in landscape drawings, distant objects appear in purple or bluish color and the same is true with respect to the color of the sky. Color pencil drawings and watercolor pencil drawings often use this kind of perspective.

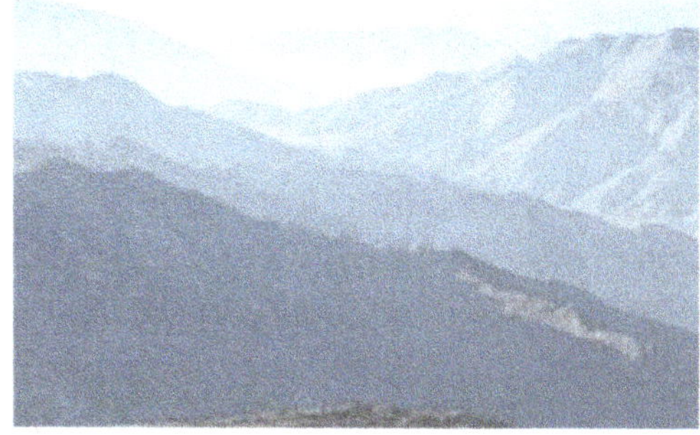

Basic Drawing Shapes

Using basic shapes will help you create drawings. Break down the object that you are trying to draw into basic shapes.

Since, these basic shapes will serve as your outline; do not make these basic shapes dark.

Making the outline light will make it easier to erase it. Sometimes, you do not even have to erase it if it is too light!

Once you have drawn the basic shapes, do not closely follow the basic shapes. If you draw following the basic shape down to the last detail, your drawing will look very rigid and amateurish.

Remember that the outline is not the product of the drawing; it is there to guide you to draw the object in a more precise.

You can deviate a little from your outline. Here are drawings using the basic shapes of a rectangle, trapezoid and circles.

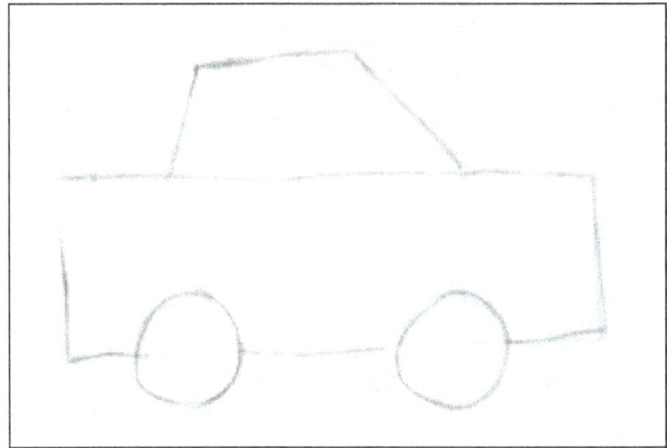

Car closely following the basic outline

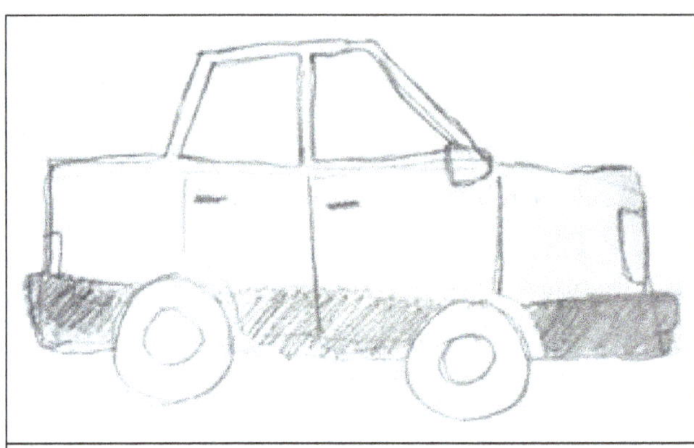

Car that did not follow the basic outline

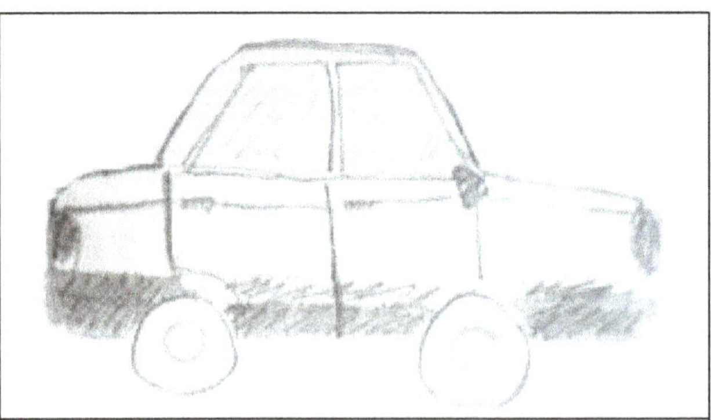

Car that did not follow the basic outline

Basic shapes do not have to be the perfect or equilateral triangle, a perfect square or a perfect circle. These shapes have variations like the triangle being isosceles or scalene, the square becoming a rectangle, diamond, trapezoid or parallelogram or the circle becoming an oval.

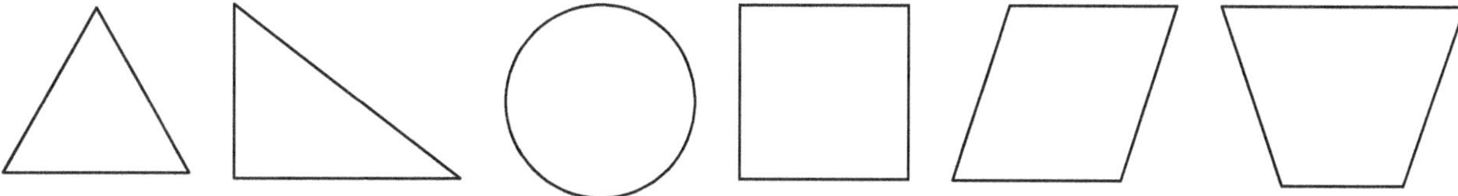

To look at three dimensional, these basic shapes can become cones or boxes.

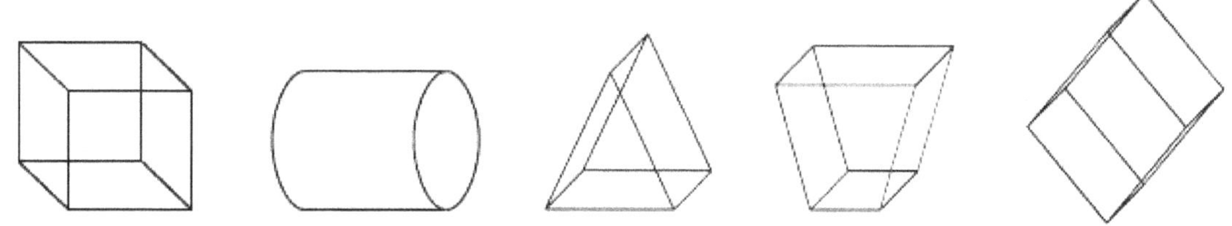

You can also use basic shapes for eyes and faces, especially with cartoons. Look at the drawings below.

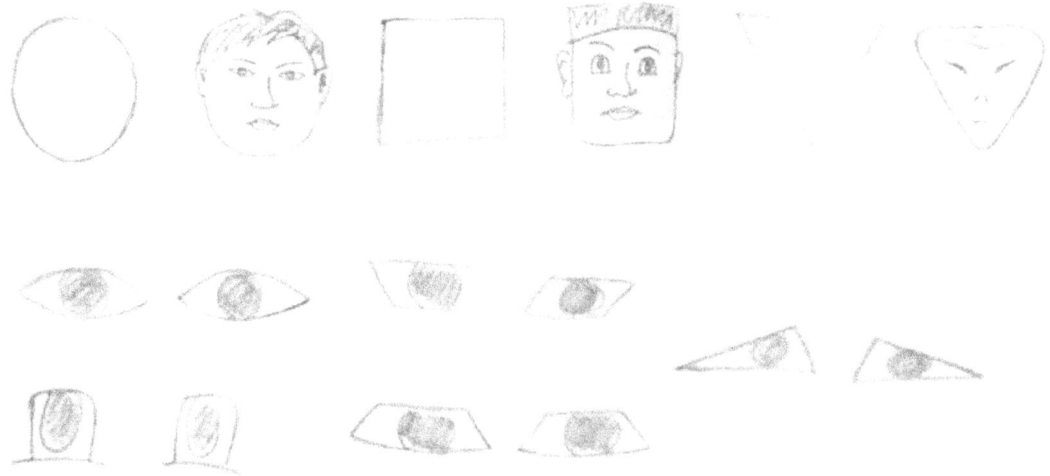

Basic Elements of Light, Shadows, and Shading

Light, Shadows and Shadow Box

Light and shadow are important in giving more life to a drawing by adding a certain atmosphere and dimension to it. In the real world, you can see a shadow in almost everything, which makes placing shadows on drawings more realistic, as putting shadows is also another way of adding perspective on an object.

This is where a shadow box finds importance since it shows how a shadow changes its direction and shapes given a particular source of light. Artists specially those drawing still objects make use of the shadow box to get a realistic picture of how light affects the direction of shadows and how to draw the proper shadings and shadows. Shadow boxes find usage in most pencil, color pencil and perspective shadow drawings.

Shading and shading techniques, on the other hand, are important to avoid your drawings from looking flat. Shading allows you to create a shape and form. Shading techniques create illusions that will make drawings more realistic.

Imagine sunlight hitting different objects, with the shadows formed; there is no even illumination on the object. The darker shades in the drawing give people the impression that it has depth while lighter areas give people the illusion of raised portion.

Translating reality to your drawings can take a lot of practice. Think of the direction of the sun and how light and shadows would interplay in the object. Parts of the object that is nearer to the sun will naturally be brighter than those that are farther away from it. Start with applying pressure on the pencil and slowly easing up the pressure to make it lighter. Gradual change in pressure will make the tones look smooth.

Constructing a Simple Shadow box

Cardboard box (20x20x16 inches)
Ruler
Pencil
Paint (white color)
Paint Brush
Double sided or heavy-duty tapes
Board cutter

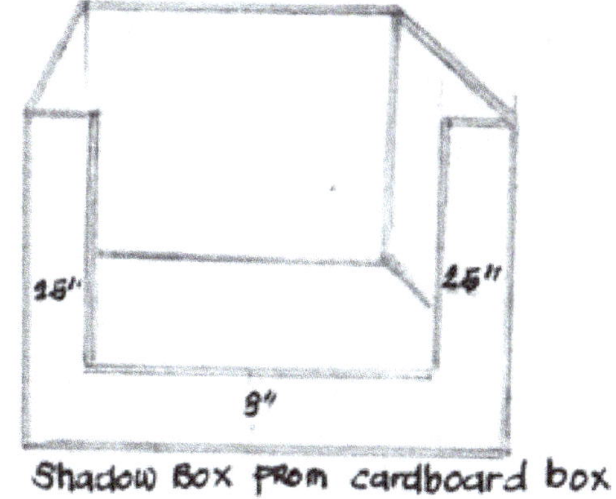

Shadow Box from cardboard box

Steps:

1. With your cutter, take away the front cover flaps of your box.

2. On the width of the box, draw a horizontal line on the bottom part at about 3 inches and on the sides draw vertical lines 2 ½ inches. Then take your cutter and start cutting out the portion as shown in the drawing.

3. Tape the bottom flaps of the box to prepare it for painting.

4. Start painting from inside the box, the three sides and the bottom with white paint preferably acrylic gesso so that it dries fast.

5. Leave under paint dries.

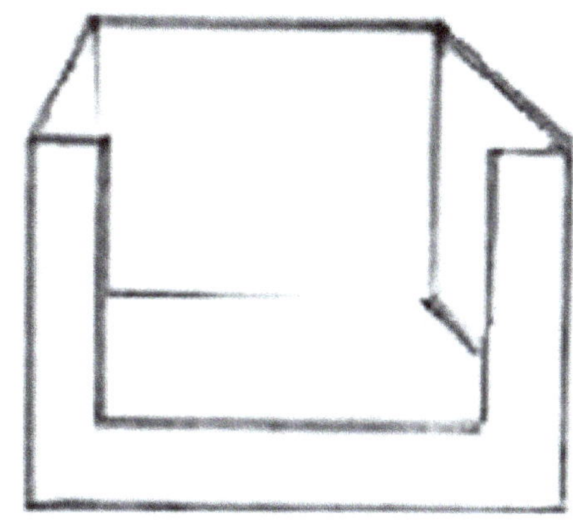

Kinds and Quality of Light

Hard Light

Hard light is that which comes from one light source that is far from the subject and this light produces both the highlights and the hard-edged shadows. Among the sources of hard light are the direct sunlight on a very bright day, the standard camera flash, the naked bulb with clear glass and many others. The use of hard light provides a moody and striking effect.

Soft light

The soft light is more subtle and creates softer shadows than hard lights. Often, household bulbs and bulbs used in photography make use of soft light as it has tendency to hide imperfections. In drawing, soft lights create regular shadows.

Basic Elements of Shading

The Highlight or Full Light

There is full light when the light hits straight an object that reflects the brightest areas in your drawing.

The Cast Shadow

This refers to those areas in the drawing, which is darkest and requires more shading. Cast shadow occurs in completely light blocked places where objects inside cast shadows on other surfaces.

The Halftone

It in addition, called the midtone that refers to those areas just in between the dark and light areas and appears to have gray value.

Midtone reflects the true consistency in terms of an object's color.

The Reflected Light

This light is near white or very light gray but not totally white and is important in projecting a dimension on an object. This is the light reflected to an object from its nearby surfaces.

The Shadow Edge

Those areas found just in between the midtone and the reflected light.

Different Shading Techniques

Regular Shading

Moving the pencil from left to right or top to bottom produces this shading technique.

Irregular Shading

As opposed to the regular shading, the pencil's direction in irregular shading changes at intervals.

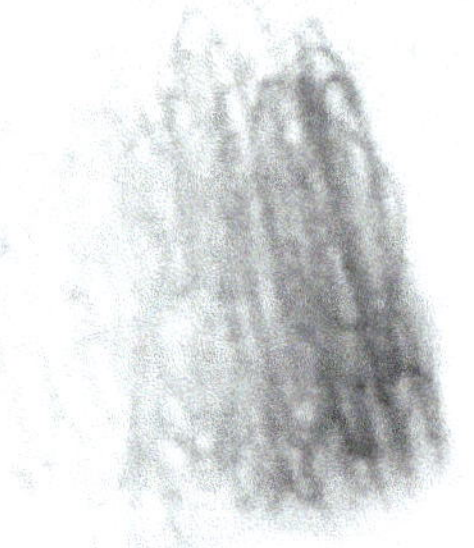

Circular Shading

Instead of making straight lines or making irregular strokes, a circular motion is used.

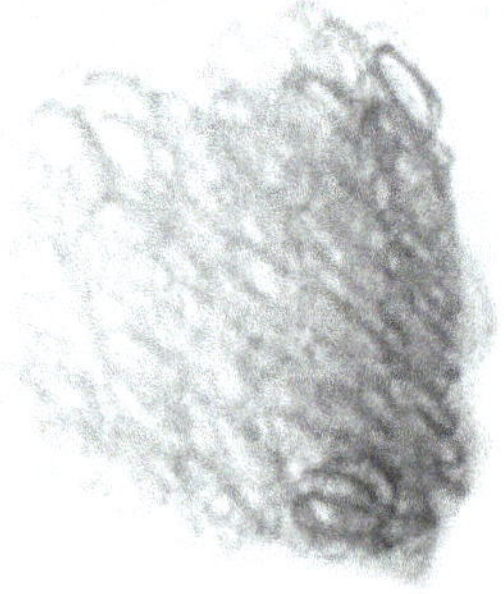

Directional Shading

Two directions are used in shading, although, they should not overlap each other.

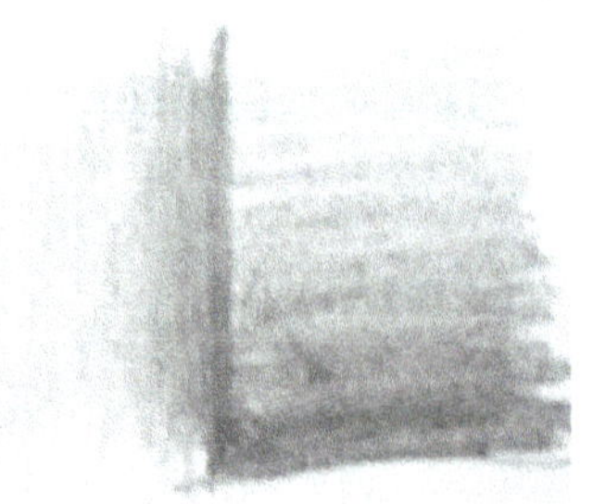

How to Add Tones and Values?

In shading, the proper lighting and shadowing is very important to make your drawings look three-dimensional and closer to reality. First, imagine the position of the light source. After this, you can determine the direction for the shadow as well as its size and shape.

Here is an illustration of a shaded pole from a certain source of light.

When you apply shading to your drawing, always remember that nothing should be all black or all white. In many drawings with shadings, the light parts contrasted against the black look white but these are very light shades of gray. There are tones in between white and black, referred to as half tones.

Below are computer-generated half tones. It is computer generated to let you see the clear distinctions between the shades.

This rectangle is the outcome of putting together all the half tones.

These half tones give out the illusion of shape. Notice how shadings can substantially change a drawing. Notice how drawings on the left look bland and boring as it against the drawings on the right.

Some Tips on Tones and Values

In shading, it is better to use a blunt pencil than a very sharp one.

Use B pencils since they are darker and will give enough depth in your drawings. Once you start shading, pressure applied changes gradually since the shading must appear or look even.

Make your strokes near to each other that it will be difficult to notice the different strokes.

Do not over shade the drawing because it will be easier to add tones than remove them by eraser.

Do not draw definite outlines when you want to shade because shading makes use of light and dark to create form.

Some Examples on Shading

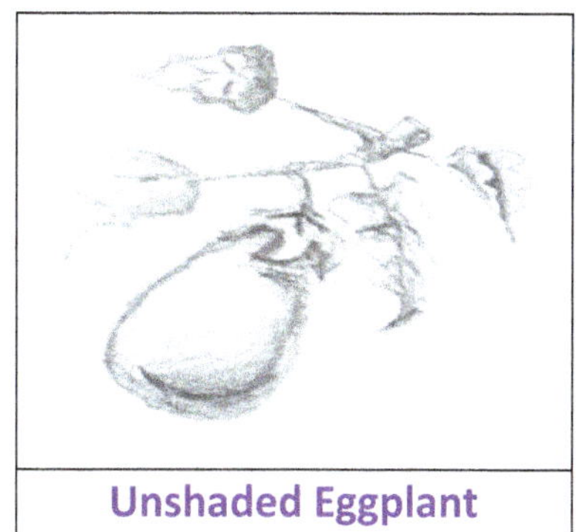

Unshaded Eggplant

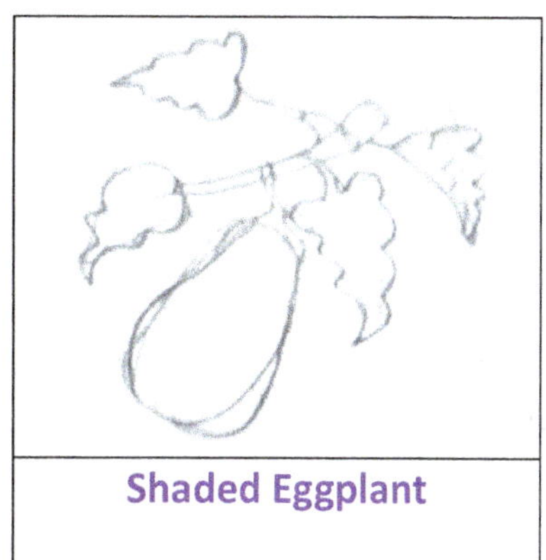

Shaded Eggplant

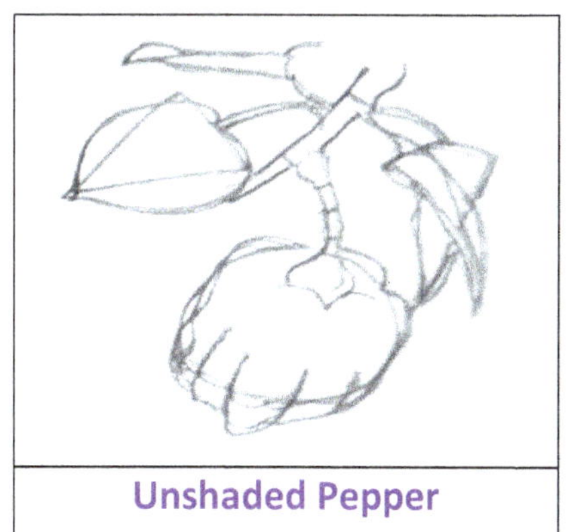

Unshaded Pepper

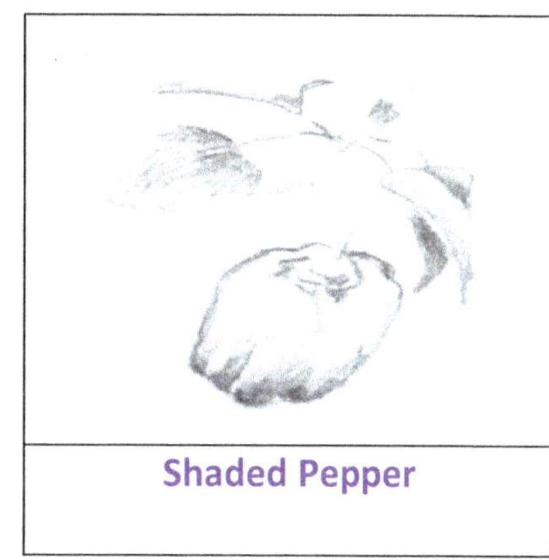

Shaded Pepper

FINISHING TOUCHES

Erasing and Dusting

Once you have completed your drawings, you need to erase our outline and excessive lines. Do not just grab any eraser available.

As once mentioned in this ebook, there are different types of erasers with different capabilities. Erasers made from pumice can erase those excessive lines but too much usage of the pumice erased in one spot will damage the paper. Sometimes, it can smear too. Pumice erasers leave a lot of residue, if it does be careful in wiping away those particles.

It is important to use brushes in wiping away eraser residue because wiping them with your bare fingers will leave oils or tend to smudge the drawings, especially if you have sweaty fingers. The brush is a great solution in wiping away those particles, or you can simply let the particles' slide out of the paper by tilting the paper and lightly shaking it.

Soft vinyl erasers are a better than pumice erasers, since they do not cause damage on the paper. Use this eraser when erasing light marks, and if you need to erase details.

Use the gum eraser if you need to erase a large part of your drawing. Since gum eraser is soft and coarse, it is not a very precise eraser. Use a brush to dust away whatever residue is present if ever it leaves residue.

Use the kneaded eraser if you need to make highlights or erase in a precise manner. You can shape this type of eraser to the form that you want it to be, thereby making it easier to erase small areas. Do not use the kneaded eraser with large areas. The kneaded eraser absorbs articles of the graphite on the paper. Once the kneaded erasure is saturated, it will not erase, instead, it will smear the paper.

Once you have erased the unnecessary lines or smudges, your drawing is now ready for the world to see and admire! Using a graphite pencil in drawing will only give you a monochromatic effect. Sometimes, that is not enough to create the picture you want to make.

Using color will bring wonders to your black and white world of pencil drawing. Colors will help you achieve the depth and visual effect that cannot attain by using a single color. In using a color pencil or watercolor pencils, the basics in pencil drawing are still used. You can still use your color pencils for shading, hatching or cross-hatching. A notable technique with a How to Add Tones and Values?
In shading, the proper lighting and shadowing is very important to make your drawings look three-dimensional and closer to reality. First, imagine the position of the light source. After this, you can determine the direction for the shadow as well as its size and

shape.

Here is an illustration of a shaded pole from a certain source of light.

When you apply shading to your drawing, always remember that nothing should be all black or all white. In many drawings with shadings, the light parts contrasted against the black look white but these are very light shades of gray. There are tones in between white and black, referred to as half tones.

Below are computer-generated half tones. It is computer generated to let you see the clear distinctions between the shades.

This rectangle is the outcome of putting together, all the half tones.

These half tones give out the illusion of shape. Notice how shadings can substantially change a drawing. Notice how drawings on the left look bland and boring as it against the drawings on the right.

Some Tips on Tones and Values

In shading, it is better to use a blunt pencil than a very sharp one.

Use B pencils since they are darker and will give enough depth in your drawings. Once you start shading, pressure applied changes gradually since the shading must appear or look even.

Make your strokes near to each other that it will be difficult to notice the different strokes.

Do not over shade the drawing because it will be easier to add tones than remove them by eraser.

Do not draw definite outlines when you want to shade because shading makes use of light and dark to create form.

Some Examples on Shading

FINISHING TOUCHES

Erasing and Dusting
Once you have completed your drawings, you need to erase our outline and excessive lines. Do not just grab any eraser available.

As once mentioned in this ebook, there are different types of erasers with different capabilities. Erasers made from pumice can erase those excessive lines but too much usage of the pumice erased in one spot will damage the paper. Sometimes, it can be smear too. Pumice erasers leave a lot of residue, if it does be careful in wiping away those particles.

It is important to use brushes in wiping away eraser residue because wiping them with your bare fingers will leave oils or tend to smudge the drawings especially if you have sweaty fingers. The brush is a great solution in wiping away those particles or you can simply let the particles slide out of the paper by tilting the paper and lightly shaking it.

Soft vinyl erasers are a better than pumice erasers since they do not cause damage on the paper. Use this eraser when erasing light marks and if you need to erase details.

Use the gum eraser if you need to erase a large part of your drawing. Since gum eraser is soft and coarse, it is not a very precise eraser. Use a brush to dust away whatever residue is present if ever it leaves residue.

Use the kneaded eraser if you need to make highlights or erase in a precise manner. You can shape this type of eraser to the form that you want it to be, thereby making it easier to erase small areas. Do not use the kneaded eraser with large areas. The kneaded eraser absorbs articles of the graphite on the paper. Once the kneaded erasure is saturated, it will not erase, instead, it will smear the paper.

Once you have erased the unnecessary lines or smudges, your drawing is now ready for the world to see and admire! Using graphite pencil in drawing will only give you a monochromatic effect. Sometimes, that is not enough to create the picture you want to make.

Using color will bring wonders to your black and white world of pencil drawing. Colors will help you achieve the depth and visual effect that cannot attain by using a single color. In using color pencil or watercolor pencils, the basics in pencil drawing are still used. You can still use your color pencils for shading, hatching or cross-hatching. A notable technique with color pencil is burnishing. It means a smooth surface will result because of layering colored pencil. Using a single color or multiple colors when employing these techniques will help create a realistic piece.color pencil is burnishing. It means a smooth surface will result because of layering colored pencil. Using a single color or multiple colors when employing these techniques will help create a realistic piece.

MIXED MEDIA APPLICATIONS

How a color pencil can bring wonders in your black and white world

Using a graphite pencil in drawing will only give you a monochromatic effect.

Sometimes, that is not enough to create the picture you want to make. Using color will bring wonders to your black and white world of pencil drawing.

Colors will help you achieve the depth and visual effect that cannot attain by using a single color.

In using a color pencil or watercolor pencils, the basics in pencil drawing are still used. You can still use your color pencils for shading, hatching or cross-hatching. A notable technique with a color pencil is burnishing.

It means a smooth surface will result because of layering colored pencil. Using a single color or multiple colors when employing these techniques will help create a realistic piece.

Since colors are involved, not only do you need to have basic knowledge on the tips and techniques on pencil drawing, a basic knowledge of the color wheel may help you in creating a great color pencil drawing.

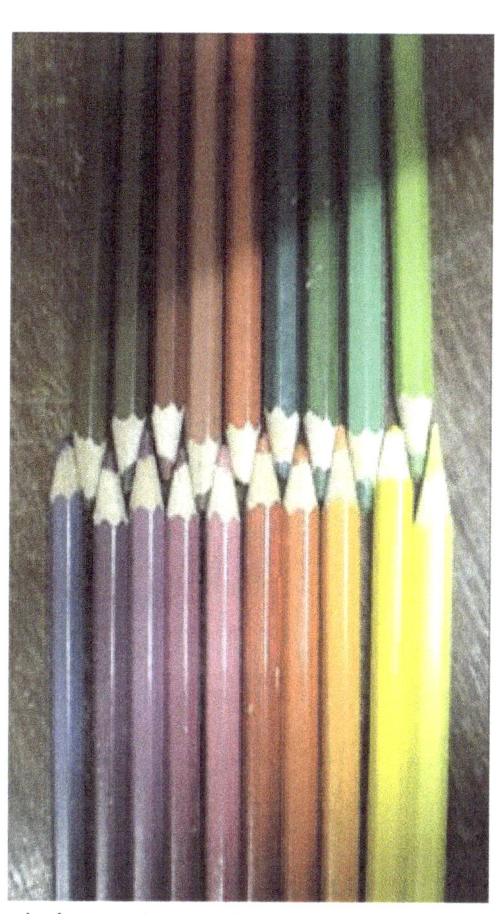

The ideal paper needed when making a drawing using color pencil is a fine-toothed paper. With this, it is easier to achieve blending and burnishing because a fine-toothed paper has less patchiness than ordinary paper.

Color pencils maybe crayon pencils or watercolor pencils. Color pencils come from wax cores. The advantage of using a crayon pencil is that it does not smudge as much.

Watercolor pencils have more or less the same effects with watercolor. Often, the usage of these pencils is for making outlines as well as for bold lines. You can saturate the pencil with water or wet a brush and start spreading color.

Look at the drawings below.

Practice Exercise:

Color Pencil Drawing - Plant

Starting with the basic shapes of
triangles or squares to help you outline
the drawing.

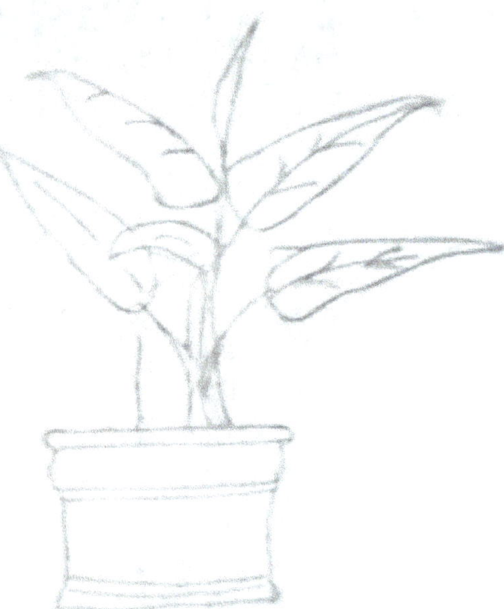

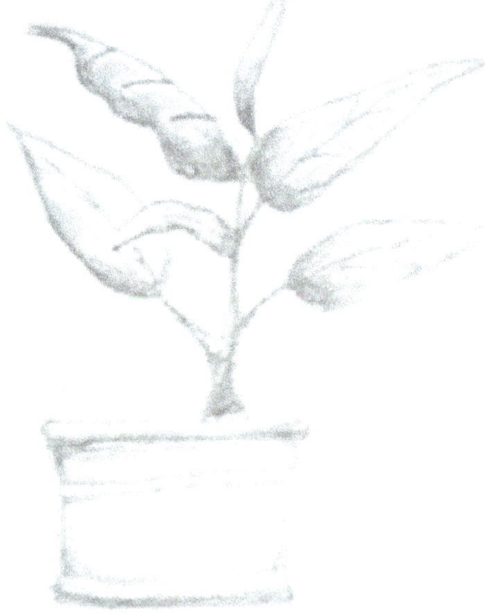

Filling in the details will make your
drawings more realistic.

Application of shading to the drawing gives it a
certain form and depth.

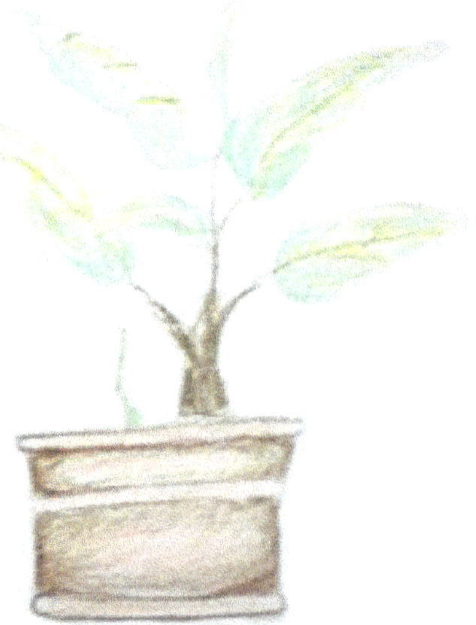

Take your drawing a step further by
adding color. Drawings look more alive
with colors.

Using Watercolor Pencils and Oil Pencils

If you are braving enough to try watercolor painting or oil painting, grabbing some of those watercolor pencils and oil pencils will give a good start on another painting medium.

The use of both watercolor pencils and oil pencils helps define your drawings.

Watercolor Pencils

Just like your ordinary pencils and colored pencils, you can sharpen and erase drawings you make from watercolor pencils. The difference lies in the not applying the dried colors in these pencils but by adding water to these pencils. The followings drawings employ some of the techniques of using watercolor pencils.

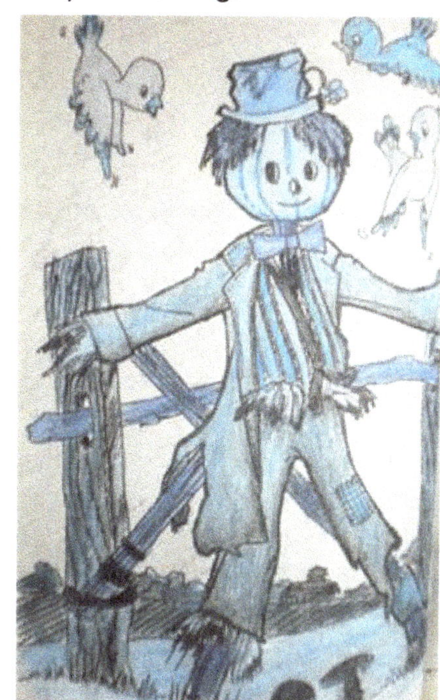

1. Wetting your watercolor pencil tips, either by dampening it with a wet brush or by dipping it in water, before drawing for you to draw intense colored lines. Later on you would see that lines become lighter and thinner as the pencil dries out; or

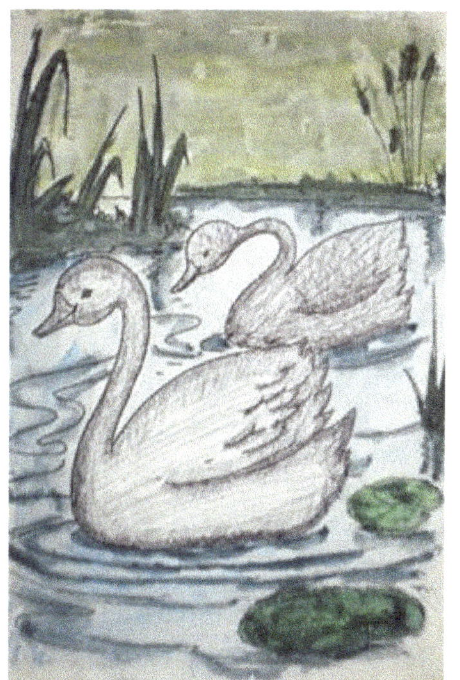

2. Brushing the drawing surface with enough water and then drawing with your pencil; or

3. Using a brush to pick up color from the pencil and then brushing on the drawing surface as the colors; or

4. Drawing on the surface with watercolor pencil and then brushing the surface with water so that water dissolves the lines of the pencil; and

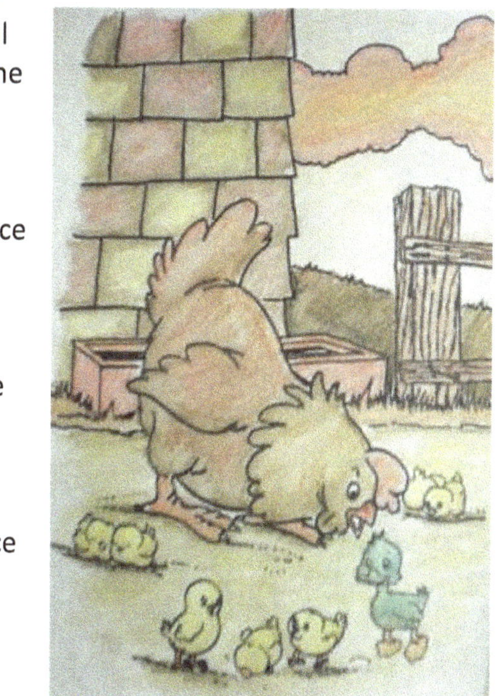

5. Scraping the colors from the pencil and sprinkling it on the drawing surface and dripping, enough water to allow it to spread.

Watercolor Pencil Drawings - Landscape

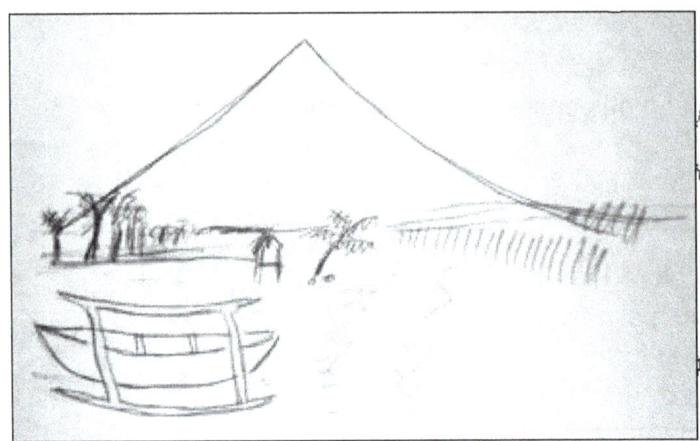

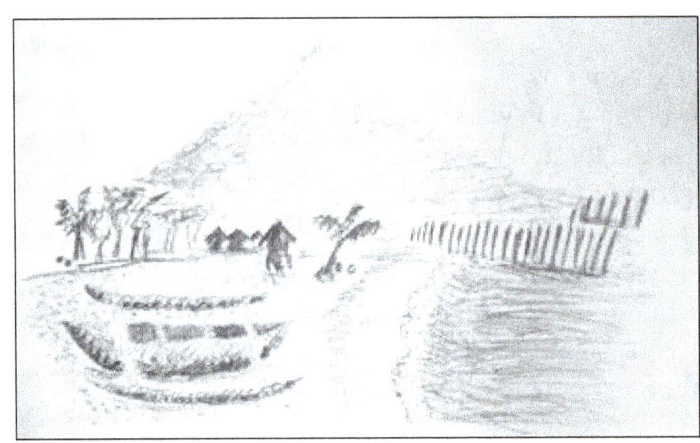

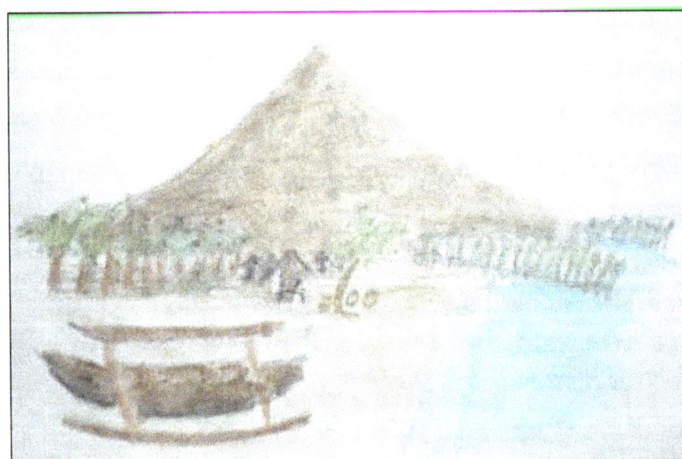

Practice Exercise:
Watercolor Pencil Drawings – Forest

STEP 1
Draw the forest's outline.

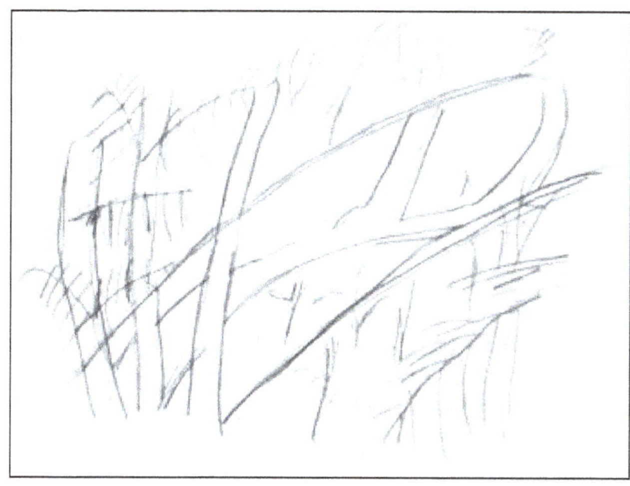

STEP 2
Fill in the details of the trees and shade the dark areas of the forest.

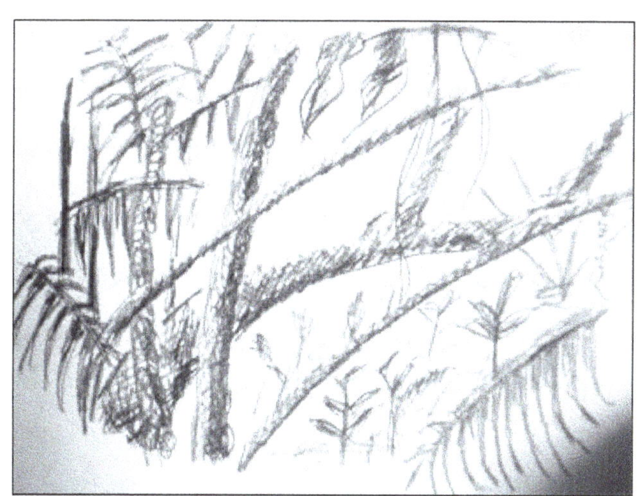

STEP 3
Enhance your forest with colors using your watercolor pencils.

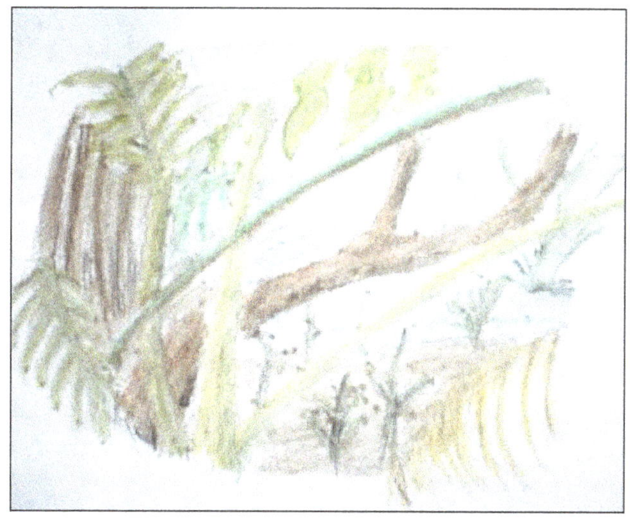

Oil Colored Pencils

Oil colored pencils are among the new coloring materials of kids nowadays since these are not messy to handle just like colored pencils. Oil Colored pencils are perfect for art works that require some highlight and shading to do.

These pencils are interesting as well because of the wide range of surfaces that one can use such as paper Mache, canvas, and fabrics and even as fillers of scratch surfaces on your furniture.

You can use the oil colored pencil by drawing in lines or in circular motion, taking note that the intensity of color produced depends on the pressure you apply on the pencil.

Today, these pencils are available in a variety of shades that you need not worry of the intensity since you can choose from a number of pencils shades that range from light to dark.

One of the characteristics of oil colored pencils is that colors are easy to blend. Just like pencils, you can correct errors by simply sanding off the surface with wrong applications.

After making your drawing, apply a sealer or a fixative to seal the drawing and make sure you are not applying to varnish over this oil colored pencils since varnish will cause the colors to spread around and smear on the surface.

Drawing with Pencils in Oil Painting

Oil painting refers to draw with the application of oil based paints to drawing surfaces. Traditionally, oil painting has its roots in early European arts that became more acceptable than its early counterparts of wax and water- soluble media.

Due to its oil base, oil paintings take time to dry and some pieces even take months to dry. The positive aspect though is that oil paintings have very slight difference in colors from the time these were still wet up to the time they dry up.

However, not unless you are a skilled painter in oil-based paintings, or you are an abstract artist; you may need to sharpen first your pencil drawing skills before you become a full-grown oil painter.

Unknown to many, even skilled oil painters have to do some drawing and sketching with their pencils and charcoals before they actually paint.

Traditionally, oil painting procedures make use of sketching first the design thereby making sure that once they start oil painting, they know that the image as they want it to appear on the canvas or on the surface is well in place.

In fact, some artists make use of shading techniques in order to highlight portions in the drawing that oil paints are not able to do.

Pencils are preferred though as compared to charcoal, as pencil marks are easier to cover. As a tip though, make sure that after making the sketch with a pencil, seal the canvas by using a clear acrylic paint to prevent smearing once you apply the paint.

Pen and Ink Drawing

Practice Exercise:
Drawing Count Von Roo

Step1: Draw one big circle for count Von Roo's head and three small circles for the ears and the nose. Draw the cross lines for the face and the outline of his cape. Draw the curve lines of the ears, cheeks and nose. Lastly, draw the outline of his right arm.

Artists may use a pen and ink in drawing. Permanency of the ink is its strength as well as its weakness.

In using a pen and ink, one has to be very precise with the strokes because unlike pencils where you can make erasures, ink is permanent.

These drawings will last for many years to come.

Drawings in ink are darker than the ones done with pencils hence ink drawings look more attractive to the eyes.

One cannot use the ink for outline or rough sketch of a drawing because of its permanent and dark nature.

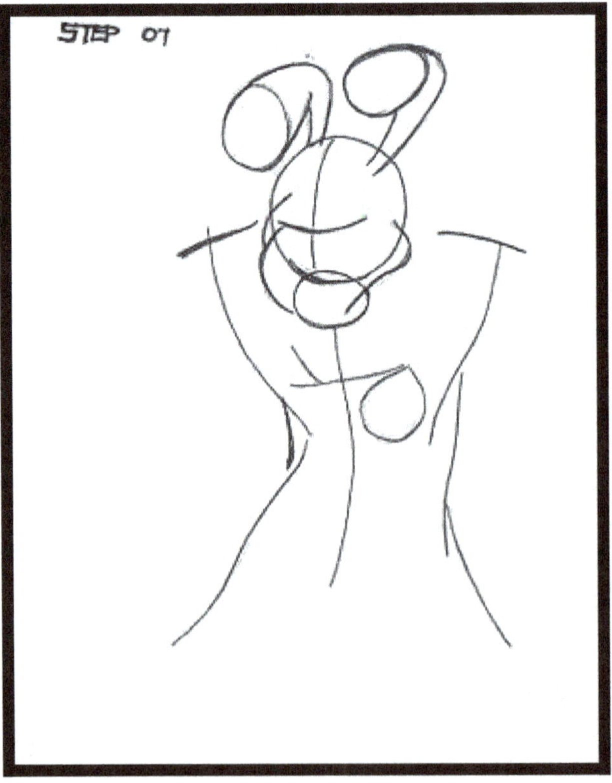

Pens involved in this technique in drawing may be dip pens or reservoir pens. Dip pens are those that you need to dip into the ink.

The nib of this type of pen is flexible and has different forms. Within the category of dip pens are reed, quill and metal pens. Inks used are drawing inks or India ink.

Reservoir pens save the user from dipping the nib of the pen to the ink. Ink constantly flows from the reservoir cartridge to the nib of the pen once the nib is moved against the paper.

The Chinese uses ink sticks, which produce works of art that lasts for many years. The Chinese grind these sticks and add water.

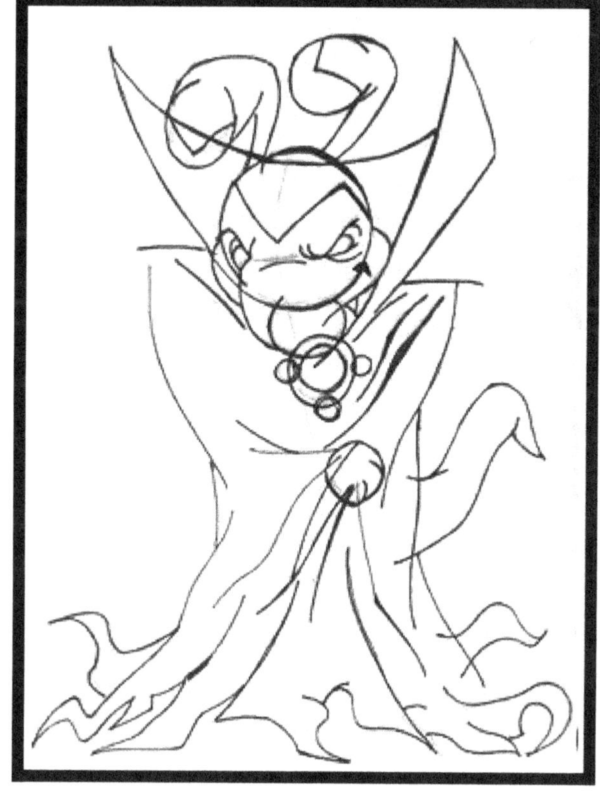

Use the 'sketching pen' because it has the best of both worlds. Its nib is not that hard as ordinary reservoir pens but it has the ink flow consistency of reservoir pens.

Some use fiber-tipped markers because they make good medium for sketching.

Pen and ink is akin to calligraphy. Artists mix calligraphy into their drawings especially with caricatures.

Many are able to make caricatures entirely in ink while some comic strips incorporate ink as part of its medium.

For many people, a pen may just be a writing instrument but for artists, a pen is a tool for great art.

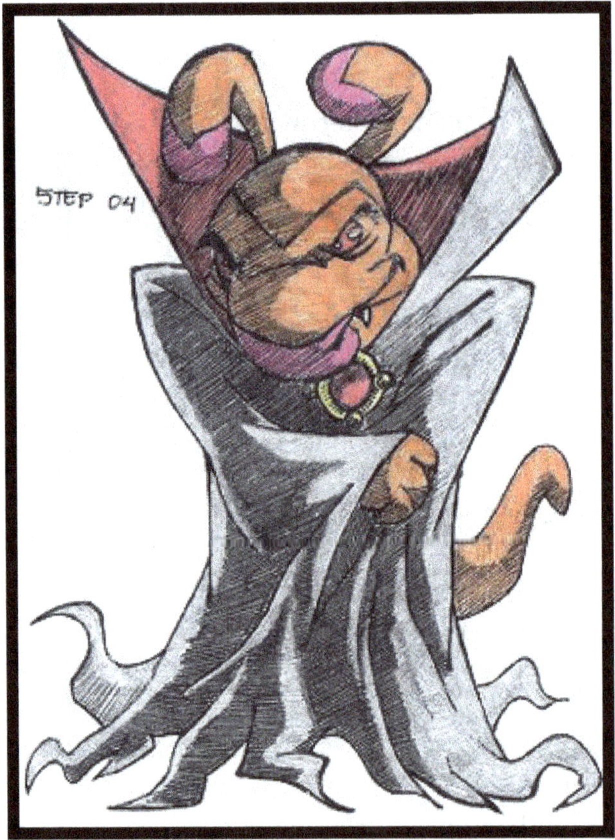

How Pencil Drawing Can Help in Your Wall Painting

Murals and wall paintings have its origins in early French and Mexican history where muralista movement rose in popularity often having used as a social tool to educate the broad mass of people with respect to political issues and trends through the art.

At present, though wall paintings nowadays are gaining popularity in individual homes transforming and adding a different mood and atmosphere to ordinary rooms and spaces in the house. Wall painting is common in nurseries, children's bedroom transformed to a fantasy world and entertainment area in the house.

One of the basic characteristics of wall painting is its large size and somehow it is both an advantage and a disadvantage since you may have a big space to work on, but on the other side, that huge space may be hard to manage as well.

In fact, once you start painting on the wall with your brushes right away, you may end up committing irreversible mistakes that require you to repaint and prepare the wall again for painting. This is where pencil drawing comes to the rescue you from getting into this trouble.

Most wall painting commissions nowadays make use of the so-called drawing grids and pencil outlines on the walls before laying their paint brushes on these walls to save paint materials, time, and to have more accuracy in their murals.

There are different ways though of making the grid lines for your wall painting. The more traditional one is by using a pencil and dividing the walls into grids and the more modern approach is with the use of projector by projecting a readymade grid unto the wall.

Most artists though prefer to use the traditional method, especially for outdoor wall paintings since the use of projector is quite expensive for long hours of use in a large painting endeavor. Some use the projector, though to make the drawing by pencil on the wall much easier and faster.

Here is a systematic method of drawing the grids for your wall painting through pencil.

1. You need to conceptualize your design and how large is. Then, you need to scale the designs into grids with the use of a ruler to show how exactly it should appear on the wall.

2. Do the same thing on the wall by measuring and marking off the corresponding grids on the wall converting the smaller scales in your image to bigger squares on the wall. Here you would need a level to check if the lines, both vertical and horizontal, are

straight enough.

3. Try to familiarize the grids of the small image with the grids on the wall and start drawing the image on the large wall with the use of a pencil making a good hand and eye coordination here.

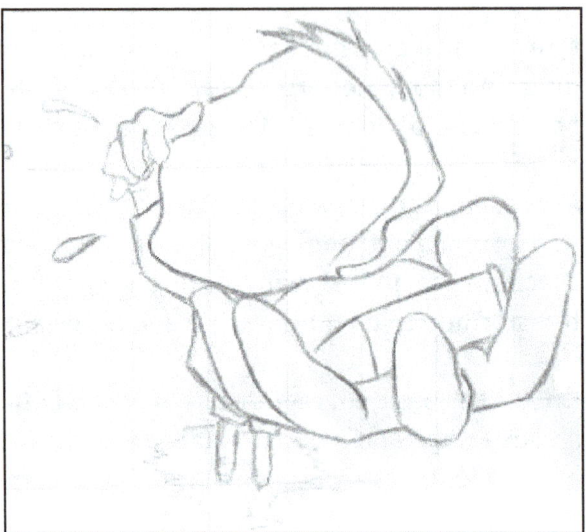

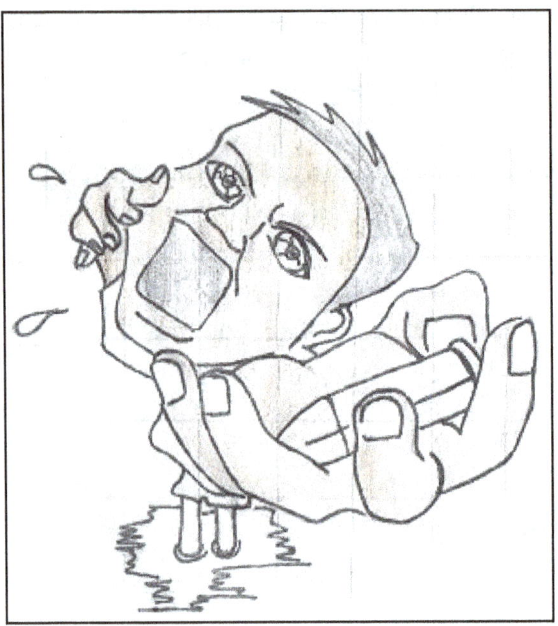

4. Once you are able to draw the image on the wall, it is now time to erase the grid lines and let the pencil drawing remain as your painting guide.

5. Paint your wall design with a fast drying acrylic paint and a flat brush making sure that you paint first the large areas. Do the detailed and smaller areas with the use of a round brush.

6. Leave the wall painted wall until it dries and then start erasing the pencil drawings.

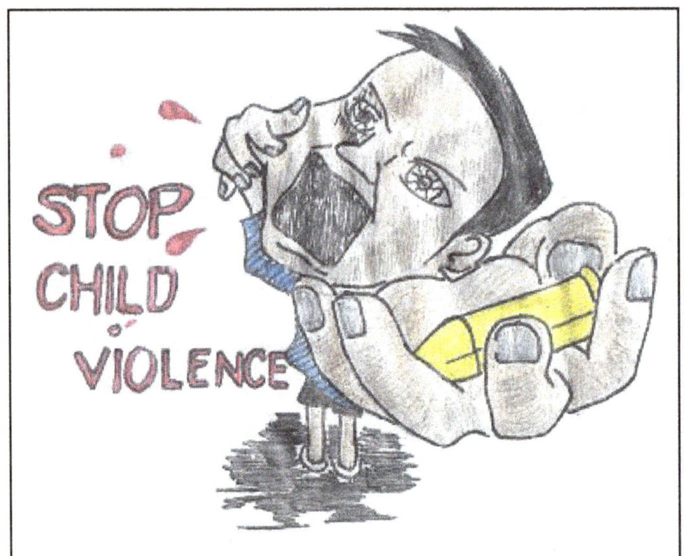

Cartoon Drawing

Cartooning and Cartoon drawing refers to drawings that often have humorous and even satirical connotations printed on various media such as newspapers, magazines, and even the world of film.

The word cartoon has an Italian origin from the word 'cartone' meaning pasteboard or the preparatory sketch for images. The cartoni or the sketches were part of early Renaissance mural arts, which served as drawings on paper used as guides for the frescoed drawings.

Cartoons may refer to animated cartoons, comic cartoons, editorial cartoons, gag cartoon and the illustrative cartoons.

Cartoon Drawing Techniques

The creation of modern cartoons starts with the cartoonists making sketches with the use of pencils, pens or computer software.

For not so technically perceptive individuals, it is better to use pencils because it is easier to make corrections without producing as many scratch papers as the pen.

Just like pencil drawing, pencil line drawing techniques and shading techniques are of great help.

Crosshatching is one of the most common techniques used to create the light and shadows of the cartoon drawing. Pointillism, on the other hand, is another way of shading areas in these cartoons.

Thus, mastery of basic pencil drawing technique is your tool to becoming the next famous cartoonist.

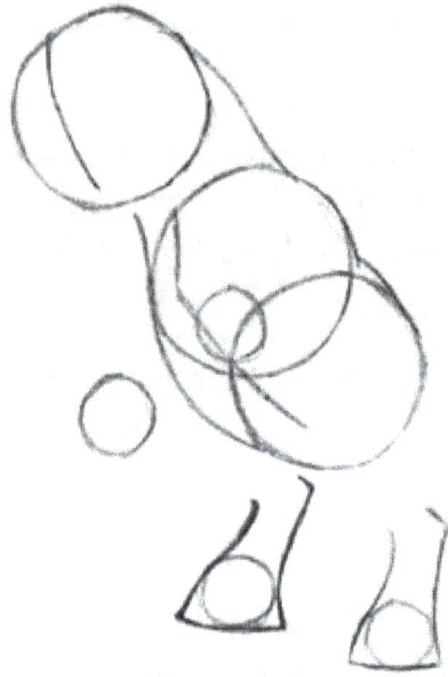

STEP 1: Draw three big circles and make two smaller circles. Draw the connecting guidelines between the large circles.
Then connect the head and body with curved lines.
Start drawing the outline of the legs. Make two smaller circles for the hooves of Uni.

STEP 2: Make the cone, the mouth, the snout and the cheek of Uni. On the earlier circles for the hooves, draw two more circles. Draw the curves of the legs. Draw two leaves shape for Uni's ears and make his wings. Draw the details of the eyes and nostrils.

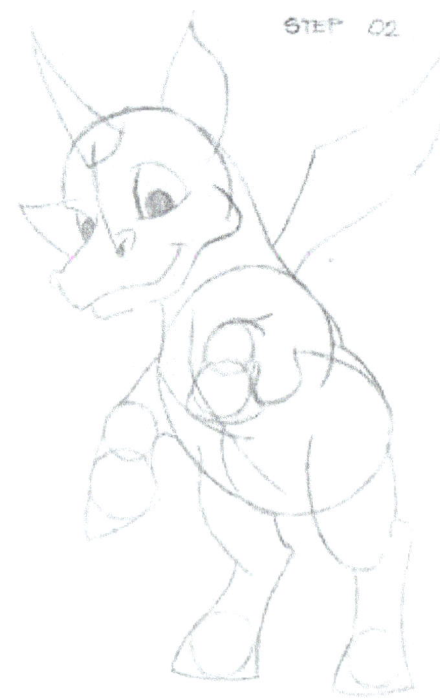

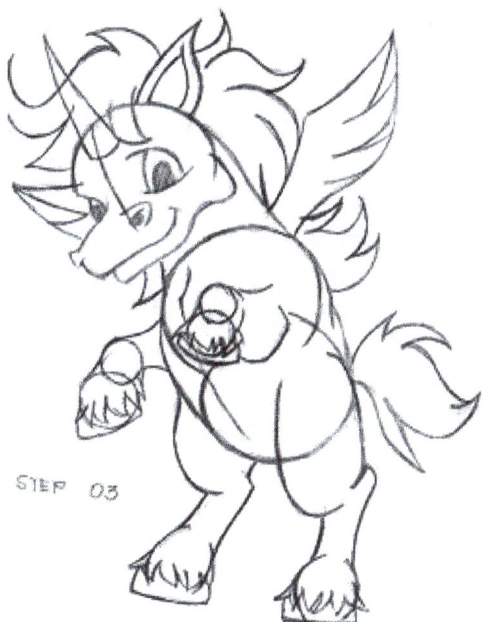

STEP 03

STEP 3: Draw Uni's tail and floppy mane and make the wings, tail and hair look feathery…

STEP 4: Finish your Uni by adding colors and erasing extra pencil marks.

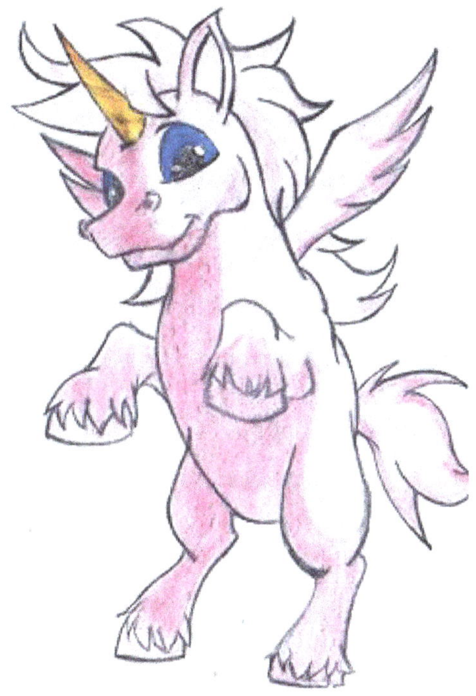

Tips on How to Draw Faster

Drawing is so enjoyable that it is easy to forget time. However, does it take you hours to complete one simple drawing?

It takes you hours, not because you are enjoying the whole process, but because you are struggling to complete a drawing?

There are artists who can draw at lightning speed and have the very good results. The key to their ability is practice. As the first tip, try giving yourself a time limit and draw as much as you can.

Do not concern yourself much with the details since the main goal is to produce as much as you can.

While you are out waiting, try to drawing objects and people moving around. Sketching, with basic shapes, helps you fill in the details later.

Remember that in drawing fast, you are not trying to replicate everything, but you are trying to make something that resembles the object you are trying to draw.

Once you got the basic shapes, you can easily put the details when you have more time.

Practice everyday and increase your capacity. For example, in the first week, you may just be able to draw basic shapes in 30 seconds.

In the second week, try to force yourself to add more details within 30 seconds. If you want to make a detailed and precise drawing, then it is best to take your time with it.

At first, everything will look worse than a child's drawing but overtime, you will learn to pace yourself and improve your technique.

CONCLUSION

Discussed and illustrated in this ebook are the basics of pencil drawing. Do not stop with this ebook.

Hone your skills by doing practice exercises.

Practice as much as you can and, let your creativity take control.

Draw everyday objects as well as your surreal visions and see what beauty you can create.

One can do so many things with one's skills in drawing.

One can draw to pass the time, tell stories, or make others happy.

Some make a living by selling their drawings.

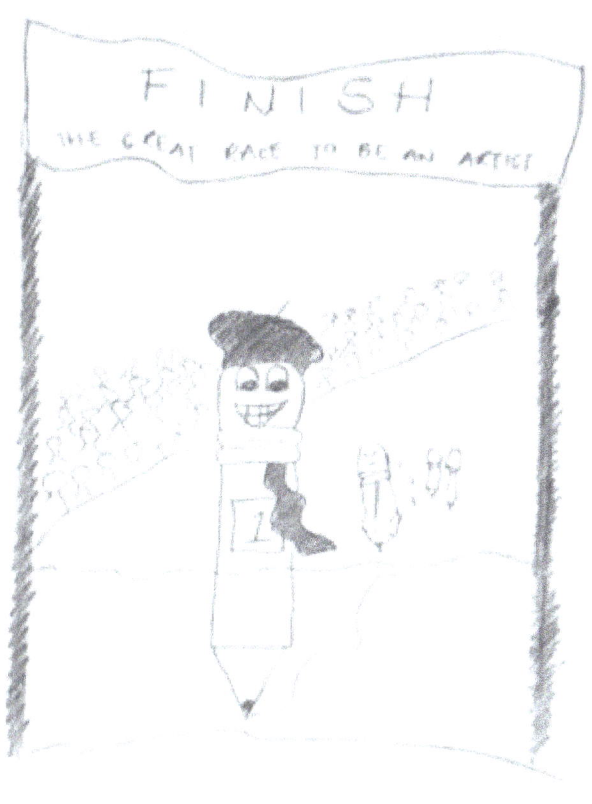

Once you have grappled with the basics of pencil drawing, expand your skills by venturing into other fields in visual arts such as watercolor, charcoal and oil painting.

It will equally be as fun and challenging as with pencil drawing.